ONE WEEK LOAN

Defects and Deterioration in Buildings
2nd edition

Books by the same author

Wood in Construction (1976) Construction Press, London.

Wood Preservation (1978) Construction Press, London.
Second edition (1993) E. & F. N. Spon, London.

Remedial Treatment of Buildings (1980) Construction Press, London.
Second edition (1995) Butterworth-Heinemann, Oxford.

DEFECTS AND DETERIORATION IN BUILDINGS

2ND EDITION

Barry A. Richardson

London and New York

First published 1991 by E. & F. N. Spon
Second edition 2001 by Spon Press

11 New Fetter Lane, London EC4P 4EE

Simultaneously published in the USA and Canada
by Spon Press
29 West 35th Street, New York, NY 10001

Spon Press is an imprint of the Taylor & Francis Group

© 1991, 2001 Barry A. Richardson

Typeset in 11/13 Imprint by Wearset, Boldon, Tyne and Wear
Printed and bound in Great Britain by TJ International Ltd, Padstow, Cornwall

The publisher makes no representation, express or implied, with regard to the accuracy of the information contained in this book and cannot accept any legal responsibility or liability for any errors or omissions that may be made.

British Library Cataloguing in Publication Data
A catalogue record for this book is available from the British Library

Library of Congress Cataloging in Publication Data
Richardson, Barry A., 1937–
    Defects and deterioration in buildings/Barry A. Richardson. – 2nd ed.
        p. cm.
    Includes bibliographical references and index.
    ISBN 0-419-25210-X (alk. paper)
    1. Building failures. 2. Buildings – Protection. I. Title.

TH441 .R53 2001
690′.24–dc21

00-040055

ISBN 0-419-25210-X

To my daughters, Clare and Sue,

whose new homes provided some of the examples of building defects described in this book.

Contents

Preface

Preparing a second edition of a technical book is always interesting because the alterations that are necessary indicate the amount of progress that has been made since the preparation of the first edition. However, with this book there seem to have been few significant technical developments since I wrote the first edition about 10 years ago and alterations are now necessary mainly for other reasons, particularly changes in regulations and standards, although I am also pleased that I have been able to introduce several improvements and simplifications in the methods used to diagnose defects and deterioration in buildings. I have also expanded some of the scientific explanations as I understand from users of the first edition that they have found these interesting and helpful.

Developments tend to be discouraged today by the restrictions imposed on new materials, particularly on those which are considered to be chemicals, such as paints and preservatives. Manufacturers cannot afford the health and environmental assessments that are required before a product can be offered on the market and, as a result, very few new products are now being introduced which require these assessments. Many manufacturers concentrate instead on political campaigns to extend the life of existing products. The effect of extreme regulation is therefore to discourage the development of new safer products and to extend the life of existing products, although they may not meet current health and environmental requirements. As a scientist with 40 years' experience in the development and evaluation of new construction materials I find that these current requirements almost always frustrate any genuine attempts to develop improved and safer materials. In one sense this system has an advantage, as it ensures that we are using materials that are well established and well understood, but in the United Kingdom, subject to both national and European restrictions, we can see our construction industry stagnating whilst there are rapid and exciting changes in many other countries who are not subject to such requirements or who choose to ignore them.

When I worked first as a construction scientist we were encouraged to support the concept of international standard and approval schemes which would allow transfer of products across national borders free from unnecessary technical restriction, but today it is clear that the critical word in this phrase is *unnecessary*, as many countries have considered it necessary to introduce new national requirements additional to those now established internationally, so that the effect has been to complicate rather than simplify, apparently to achieve bureaucratic satisfaction rather than to improve efficiency or safety. I do not object to new regulations which improve safety, but is there any need for any other type of regulation? If controls are necessary they should be performance requirements, leaving complete freedom for the development of products and systems which will best meet both regulations and market requirements, as this is the way to encourage the development of improved and safer products. Whilst this book is not concerned directly with materials or products, it considers defects and deterioration which can be avoided in many cases by new developments. Let us hope that improvements will be encouraged in the future and that the need for books of this type will diminish, except perhaps in relation to old buildings. Unfortunately it is my experience that history repeats itself and a problem eliminated many years ago will become a new problem tomorrow. For this reason I believe that this book will always be useful in assisting the diagnosis of defects and deterioration in buildings.

Barry A. Richardson
Winchester 1999

Preface to the first edition

My family first became involved in building defects and deterioration many years ago when my father, Stanley Richardson, a pharmaceutical chemist in Winchester, was asked whether he could 'dispense' a rather strange 'prescription'. The formulation was a mixture of paradichlorobenzene, soft soap and cedar wood oil which had been developed by Professor Lefroy of Imperial College in London, and it was required in large quantities for application using French vineyard sprayers to the roof timbers in Winchester Cathedral to control Death Watch beetle infestation. My father warned the contractor purchasing the formulation of the dangers of paradichlorobenzene but he seemed unperturbed; he died several months later, apparently as a result of exposure to this volatile and dangerous chemical. The architect to Winchester Cathedral was keen to continue the Death Watch beetle eradication work and asked my father whether he could develop a reliable and safer treatment.

Following a detailed study of the Death Watch beetle my father developed a formulation based on chloronaphthalene wax and orthodichlorobenzene in trichloroethylene solvent; orthodichlorobenzene was considerably safer than paradichlorobenzene but it was later replaced by the insecticide rotenone, a natural extract of derris, and a large part of the trichloroethylene solvent was replaced by a kerosene solvent known as petroleum distillate in order to reduce the anaesthetic effect of the formulation. As the formulation was designed to eradicate Anobid beetles, it was originally named Anobol when it was first introduced in 1934 but it was subsequently renamed Wykamol after William of Wykeham, the Bishop who was responsible for so much of the restoration and reconstruction of Winchester Cathedral in which the treatment was first used.

The operatives applying the treatment were equipped with respirators which gave them a rather sinister appearance and the photogenic nature of the treatment in Winchester Cathedral soon attracted the attention of the national press. The publicity resulted in enquiries from all over Britain, and in July 1935 Richardson & Starling Limited was formed to manufacture the treatment product.

It was soon realised that Death Watch beetle is rather selective in the buildings that it infests as it needs hardwoods such as oak, which are infected to a limited extent by fungal decay, whereas the Common Furniture beetle occurs more widely as it also infests dry furniture and softwood structural timbers. However, fungal decay appeared to present even more severe problems, the ultimate cause of the fungal infections being dampness. Chemical products were developed which could greatly assist in remedying these problems, but chemical treatments alone were insufficient. The affected buildings needed to be inspected, the deteriorating organisms identified and the ultimate problems diagnosed. In theory, architects and building surveyors should undertake such tasks but few of them had sufficient knowledge and experience, and most of them lacked enthusiasm for such a specialist task. Inspections needed to be reported and remedial treatment specifications prepared to enable contractors to carry out the works, but few building contractors were actually keen to take responsibility for exposure works or to apply chemicals with which they were unfamiliar. Clearly the supply of treatment chemicals was insufficient and an integrated manufacturing, inspecting, reporting and contracting service was required, with property owners and their architects or surveyors able to select the services necessary for their particular requirements.

I was very young when I first became involved in these activities, accompanying my father on visits to buildings which were being inspected or treated, but later I worked as a treatment operative during school and university holidays. Eventually I progressed to the company laboratory and finally became the Research and Technical Service Manager. Although my laboratory was concerned with research and development into new and improved products, as well as providing quality control, safety and other routine services to the manufacturing and contracting operations, we also provided the company building inspectors, sales representatives and customers with technical service. Obviously there were few enquiries involving the normal products and services supplied by the company and most of the enquiries were concerned with unusual problems. A typical enquiry might involve identifying an insect or fungus, and reporting upon the damage that it might cause and its probable origin, but many enquiries concerned problems on site, such as diagnosing the cause of dampness or stone deterioration, problems that were often solved relatively easily through the combination of scientific knowledge of structural materials and practical experience as an operative working in buildings.

In 1965 I left Richardson & Starling Limited to become a consulting scientist specialising in the deterioration and preservation of structural materials. Initially most of the work of the practice was concerned with research and development for the chemical industry on new and improved structural material treatments but there was always a significant amount of investigative work involving site inspections and laboratory examination of samples for clients requiring advice on difficult problems or assistance with civil or criminal litigation resulting from defects in buildings.

In 1968 the practice moved to Penarth House near Winchester and became Penarth Research Centre which was operated from 1973 by Penarth Research Limited. The overseas activities expanded steadily and Penarth Research International Limited was formed in 1979 in Guernsey to handle these activities, as well as to provide marine testing facilities. The economic recession from 1980 onwards severely reduced the volume of routine industrial research and development work, and Penarth Research Centre was closed in 1985 when the site was required for a road scheme, although PRIL continues to operate in Guernsey where one of its best known activities is the organisation of an annual conference on building defects and failures which attracts delegates from all parts of the British Isles.

Over the years my own activities had also changed and I had become mainly involved in expert witness work in connection with claims and litigation, as well as arbitration in disputes where the parties require a technically knowledgeable arbitrator. These activities are more appropriate to private practice and I now practice again in my own name as I did when I first became a consulting scientist in 1965, although I remain a director and consultant to Penarth Research International Limited through which I am still involved in industrial and overseas work. This broad range of activities is important, feeding me continuously with additional knowledge and experience. My investigations are not confined to problems arising on site but often involve defects in the manufacture of structural materials. I am sometimes accused of seeing problems wherever I look because, it is suggested, I am only involved in investigating defects and failures. In fact, I continue to be involved in industrial research and development work, and in routine advisory work on the testing and selection of materials, but the importance of expert witness work lies in the thoroughness with which investigations are made, sometimes disclosing unexpected problems, perhaps prompting research and development work, or changes in regulations and advisory literature in order to avoid such problems in the future.

Friends will often say that they envy me the travel involved in my work, simply because they have heard that I have visited an interesting or exotic place. Travel is actually the most tedious feature of my work but it is fully compensated by the many interesting people that I meet and challenging problems that I encounter. For many years I have lectured to students, presented papers at conferences, and spoken to surveyors at their branch meetings, hopefully helping them to avoid the problems that I have been required to investigate but also encouraging them to attempt their own investigations. My book *Remedial Treatment of Buildings* was published in 1980 in an attempt to provide architects and surveyors with information on wood treatment, damp-proofing, masonry treatment and thermal insulation which would enable them to make their own inspections and prepare their own reports, but it immediately became apparent that a further book was necessary covering a wider range of building problems and concentrating particularly on diagnosis. This book is the result. I have tried to make it interesting by including descriptions of actual investigations rather than concentrating on theory alone, and by covering buildings of all ages and types. Whilst this book is intended to assist architects, surveyors and engineers in their normal work, it will be appreciated that the discovery of defects or deterioration is often the first stage in a process starting with a complaint and perhaps leading to a claim and a dispute that may be resolved only by arbitration or litigation. This book may therefore be equally useful to lawyers and arbitrators, and includes comments where appropriate on case law on liability.

Unfortunately the persons least likely to read this book are those who are over confident or not particularly conscientious or competent, the persons who are the cause of most of the defects that I investigate.

Barry A. Richardson
Winchester 1990

Defects and deterioration

1.1 Introduction

The purpose of this book is to assist architects, surveyors and engineers to recognise, diagnose and avoid problems in buildings, and to decide to obtain specialist assistance if appropriate. Where a problem leads to litigation or arbitration a much more critical assessment is required, usually resulting in comments on the problems in relation to the contract documents and normal good practice, as represented by British Standard specifications and codes of practice and other readily available sources of guidance. Where an action is against an architect, surveyor or engineer, an expert in the same discipline will be required to give an opinion on professional duties and the actions that a reasonably competent and diligent person should have taken in the circumstances, but the architect, surveyor or engineer employed as the expert may not necessarily be responsible for the actual investigation, which may involve unusually critical site inspections, laboratory testing and extensive knowledge of building science in order to make a reliable diagnosis of the causes of the problems. Such investigations are the province of the building scientist, usually working in close cooperation with the architects, surveyors or engineers who are also engaged to give expert evidence. It is the thoroughness of these investigations that makes them so interesting and instructive, as they frequently disclose matters which would not be considered in normal structural surveys. It is this more detailed information that is presented in this book, in the hope that it will assist in better understanding of building problems.

It is not intended that this book should be a comprehensive account of defects and deterioration encountered in buildings. Instead the book describes some of the problems that seem to present the greatest difficulties to architects, surveyors and engineers, and which are therefore most frequently referred to building scientists for investigation. The defects described arise generally through lack of care or knowledge in specification or workmanship; this book is not concerned with obvious design defects, such as inadequate beam sections in relation to design loads.

1.2 Legal aspects

Problems and failures in buildings can be broadly attributed to either defects or deterioration. Defects arise due to error or omission, that is, breach of contract or negligence by a designer or contractor, but deterioration is a natural process which may be unavoidable, although minimised by care in design and the selection of materials. However, where the rate of deterioration is excessive it may be due to a defect, such as the selection and use of unsuitable materials, or an event, such as a water leak resulting in fungal decay. A defect in such circumstances may be due to a breach of contract or negligence during construction or repair, or it may be a failure by a building owner or tenant to maintain the building adequately or to repair it promptly after accidental or weather damage.

When a problem or failure arises and a building owner or tenant decides to seek compensation, it may be very difficult to establish that a breach of contract or negligence has occurred, rather than normal deterioration, lack of maintenance or failure to repair promptly. One difficulty is that a cause of action can only arise if a person or firm has suffered an actual loss. The implications are best illustrated by an example, perhaps of a building with inadequate foundations. If the faulty foundations are discovered soon after construction, the owner can claim against the contractor for breach of contract if the contractor failed to observe the design requirements which formed part of the contract. However, litigation must commence within the statutory limitation period, that is within 6 years of practical completion for a normal building contract, but within 12 years for a speciality contract, that is a contract made by deed or under seal. Generally, a building owner has retained an architect to design the building and supervise construction, so that a contractor might claim in defence that the inadequate foundations were installed to the satisfaction of the architect. A building owner may therefore claim also against the architect for breach of contract for failing to supervise construction properly, or alternatively for failure to design the foundations properly, if it is found that the contractor actually followed the design. The situation may be further complicated if the relevant work was subcontracted; an architect might subcontract or assign foundation design work, to an engineer, and a contractor might use a subcontractor for specialist foundation work, such as piling.

In fact, inadequate foundations may not be detected until much later when some event occurs which leads to investigations, such as the development of a settlement crack, often well beyond the period when an action can be brought for breach of contract. However, an action for a latent defect can still be brought in tort, that is a claim for a wrong arising through negligence, as the limitation period is then 6 years from the date on which the damage occurs or, if it is later, 3 years from the date on which the damage was first reasonably observable, with an ultimate limit of 15 years from the date on which negligence occurred, that is usually the date of practical completion. In the case of architects or engineers it is often suggested that limitation should run from the date of completing designs, but if they are supervising construction they are also responsible for reviewing their designs and specifications until the date of cessation of their involvement in the project, usually the date of issue of the final certificate, or even later if they renew their involvement in the project, perhaps through being asked to advise on a problem that has occurred and which is subsequently found to be associated with a defect. These comments on limitation are included to illustrate the way in which lawyers and courts have become dependent on expert witnesses for guidance on limitation matters, as only an investigating expert can advise authoritatively on the dates when damage occurred or was reasonably observable. These comments on limitation are based only on the law of England and Wales as represented by the Limitation Act 1980 as amended by the Latent Damage Act 1986, although similar limitation legislation applies in all states of the British Isles.

Whilst an action for breach of contract can only be brought between contracting parties, an action for negligence can be brought by anyone who is likely to be affected by a negligent act, so that a tenant or user of a building can commence an action, as can a person passing in the street who is injured, for example, by falling masonry. It may seem unfair that an action for negligence cannot be brought until there is a cause of action, that is actual loss or damage, but this seems to be a deliberate decision of English law to discourage unnecessary actions; it is recognised that it is very rare for buildings to be erected strictly in accordance with their designs and current good practice, but it is equally rare for buildings to suffer damage as a result. An obvious example would be a situation where a contractor was unable to obtain the specified roofing tiles and used instead the nearest available equivalent. Obviously the substitution was a breach of contract, but it would not be negligent, even if carried out without approval, unless it resulted in actual loss or damage.

The only exceptions to these general rules arise when a structure is *doomed to fail*, as in the case of an overloaded component which is progressively distorting, or when there is an imminent threat to health and safety because part of a building is certain to collapse if subjected to wind or snow loads of normally encountered severity, or there is a danger of rapid spread of fire. In such circumstances a cause of action arises because the defect must be corrected immediately, involving a financial loss; indeed, it is necessary to remedy the defect promptly in order to mitigate the loss and to avoid the much greater claim that might arise should a catastrophe occur, such as a structural collapse or fire causing death or injury.

Compensation for defects can be valued in various ways. It may seem obvious that the compensation should equal the cost of repairs and any consequential losses, but a repaired building may never be the same as a properly constructed building, and an alternative method of valuation would be the diminution in the value of the property due to the defects. Diminution in value usually applies to claims arising through negligence in assessing the condition or value of a property prior to purchase, but diminution is difficult to quantify. Obviously, diminution in value is directly related to the cost of necessary works, but the relationship depends on the state of the market when the offer based on the negligent assessment was accepted by the vendor. If the market is buoyant and there is a strong demand for such properties at the asking price, prices will be rising and a prospective purchaser will be reluctant to abandon a purchase and search for an alternative similar property at a relatively higher price. If a survey discloses necessary works, the prospective purchaser will probably be willing to tolerate the inconvenience and perhaps part of the cost of the works. The prospective purchaser is therefore likely to seek a reduction equal to the estimated cost of the necessary works, but may be forced to agree a reduction to only perhaps a half or three quarters of the cost. However, if the market is sluggish with weak demand and static or falling prices, a prospective purchaser would have little to lose by abandoning the purchase and seeking an alternative property, and would not therefore be willing to tolerate either the cost or inconvenience of necessary works. In such circumstances, the purchase would be unlikely to proceed unless the vendor was willing to accept a substantial reduction, such as 100–150% of the estimated cost of the works. Another method of valuation, appropriate to commercial property although rarely used, relies on the loss of profit that has resulted when a building has proved to be unsuitable for the purpose for which it was intended.

These comments on legal liability have been included to emphasise the seriousness of problems and failures in buildings due to errors and omissions, but such problems are not confined to new construction. Obviously, identical problems apply to repair works of all types, particularly perhaps remedial treatments involving damp-proofing, thermal insulation and timber preservation; these introduce their own special problems as they generally involve specialist contractors, who also provide inspection, report and specification services so that they assume professional responsibility as well as normal contractual responsibility for their works. Their professional responsibilities involve the assessment of existing defects and their preparation of specifications for remedial works, but often their approach is too narrow, perhaps overlooking the fundamental causes of the problem that they are investigating. For example, in the case of fungal decay in a modern timber frame building, they may suggest remedial preservation treatment but fail to comment on defects in the internal vapour barrier or external ventilation which have resulted in interstitial condensation causing dampness and consequential fungal decay. Some remedial treatment contractors believe that they are not liable for the professional thoroughness of their inspections and the accuracy of their reports because they do not charge fees for these services; this is not correct. It is well understood that the costs will be incorporated into any charge for remedial works, and some contractors have discovered that they have the same liability for any inspections that they make and reports that they submit as a chartered surveyor or architect. In fact, a contractor is only usually liable for errors and omissions in his inspection and report if it can be shown that he acted negligently; a contract, if one exists, is usually restricted to making an inspection for the purpose of preparing a specification for remedial works which the contractor considers to be appropriate. A contractor can minimise liability by referring in advertisements, literature and letterheads to inspections for preparing estimates, rather than surveys and reports which imply an authoritative and impartial assessment.

The obvious solution is for property owners to rely only on a proper independent professional assessment, but the arguments are not necessarily convincing. An architect or building surveyor will charge for his services, whereas specialist remedial treatment contractors will usually offer to provide survey and report services free of charge. In fact, specialist contractors attempt to minimise their inspection and report costs by limiting their investigations to the remedial works that they would like to carry out, incorporating the inspection and reporting costs in their estimates or quotations for these works. In addition, an architect or surveyor may be reluctant to give an opinion on specialised matters, such as dampness and timber decay, and will recommend in any case that a specialist contractor should be engaged to inspect, report and to carry out necessary remedial works! The situation is thus thoroughly unsatisfactory: architects and surveyors often failing to provide a comprehensive assessment of a building, while specialist remedial treatment contractors are expected to provide professional services although they are, in fact, only contractors inspecting buildings to prepare estimates and quotations for works that they would like to carry out. This situation has been clarified by a case in which a chartered surveyor was found to be negligent for detecting damp conditions likely to represent a fungal decay risk and for failing to make further investigations; he recommended that a remedial treatment specialist contractor should be consulted, but he failed to assess the situation himself.

The most serious situations probably arise in connection with surveys prior to purchase in which the status of the survey is of crucial significance. Where a survey is for valuation purposes alone, it will not necessarily involve any reference to the condition of the building. In an extreme case, the valuation may be based on the site value alone and the building may be ignored, but where the building itself has significant value a valuation surveyor may report structural defects and the need for further investigations, repairs or maintenance where these aspects affect the valuation; this does not necessarily mean that the building is free from other defects which may require attention. A very strange situation arises where a prospective purchaser applies for a mortgage loan and is required by a bank or building society to pay for the valuation survey and report, but does not receive a copy. The legal interpretation of the situation seems to be that the prospective purchaser does not have a contract with the surveyor, but the courts have indicated that a prospective purchaser is entitled to rely on the valuation information and may be successful in a legal action alleging negligence against the surveyor if the valuation subsequently proves to be incorrect. In recent years the Royal Institution of Chartered Surveyors has recommended that surveyors should offer, whilst making a valuation survey for mortgage purposes, to also provide the prospective purchaser with a basic *House Buyer's Report* to a standard format, but a report in this form is still very limited in scope compared with a proper structural survey. Clearly an architect or surveyor has a duty of care when carrying out surveys of all types but, when a problem arises, they can only be considered in breach of contract or negligent if it can be shown that the defect should have been discovered by a reasonably competent and diligent surveyor; the most important point is to detail in all reports the areas where access was inadequate and those where further investigations are necessary.

1.3 Investigations

A structural survey is a routine assessment of the condition of the building which is intended to detect any defects or deterioration requiring attention. Structural surveys are sometimes arranged because the owner or tenant of a building suspects a problem, although most structural surveys are for prospective purchasers. Structural surveys are normally carried out by building surveyors or architects; a building surveyor should be appropriately qualified, such as a member of the Royal Institution of the Chartered Surveyors in the building surveying division, rather than in one of the other divisions, such as general practice, quantity surveying or valuation surveying. Guidance on structural surveys is available from various sources such as the book by J. T. Boyer, *Guide to Domestic Building Surveys*, and the RICS booklet, *Structural surveys of residential property*, originally issued as a Practice Note in 1981 but described instead as a Guidance Note when a revised edition was issued in 1985. The RICS note emphasises the need for the surveyor to confirm the client's instructions and to appreciate the client's requirements. Obviously, a reliable and realistic assessment of the structural condition of the property is required, although it may be reported that defects or deterioration are possible rather than established, and that further investigations are necessary or at least advisable. This is a common difficulty with surveys for prospective purchasers: the permission of the vendor must be obtained before carrying out exposure works, yet the prospective purchaser requires to know the scope and estimated cost of both essential and advisable work, so that this can be taken into account when making an offer to purchase the property.

A survey report may advise that further investigation should be carried out by another professional person, such as a structural engineer, or that a specialist contractor should be asked to inspect and prepare a specification and estimate for appropriate work. Remedial wood preservation and damp-proofing treatment contractors are often invited to prepare specifications and estimates in this way, but so are many other contractors involved, for example, in plumbing and heating engineering, piling and roofing. There is a temptation for the building surveyor or architect responsible for a structural survey to suggest that these contractors should be instructed to 'survey and report', but this is not their proper function; they should be instructed to inspect only to enable them to prepare specifications and estimates for work defined by the building surveyor or architect.

Investigations into defects or deterioration in connection with claims or disputes involve a rather different approach, focusing on a particular problem, in contrast to a normal structural survey which is designed to assess the general condition of a property and detect any matters requiring further investigation. The differences between defects and deterioration have already been explained, but only defects can support a claim or dispute, in the sense that deterioration is a natural process which is to be expected. However, when deterioration is observed it must be established whether it is occurring unexpectedly rapidly through some error or omission in specification or construction, or through inadequate maintenance, or through a defect.

A realistic approach to deterioration is perhaps best illustrated by an example.

Bituminous felt is often used as a roof covering for permanent buildings, even though the felt has an effective life which is much less than the design life of the building. Damp stains on ceilings beneath the roof will suggest failure of the felt covering. This failure may be due to natural deterioration, such as loss of volatile components causing brittleness and cracking of the felt; the resultant leaks are then attributable to neglect to maintain the roof properly by replacing the felt at appropriate intervals. However, premature development of leaks may be due to poor workmanship or inadequate materials, in the sense that an unsuitable grade of felt or insufficient layers were used. Alternatively, some accidental event might have occurred, such as dropping an item on the roof which penetrated through the felt covering. Localised cracking in felt may also be due to thermal movement in the building which may overstress the covering as the felt ages and becomes more brittle. All these failures involve rainwater leakage through the roof covering, but dampness may also be caused by interstitial condensation beneath the felt covering, a defect in design or workmanship which is described in greater detail in section 4.7. However, whether the dampness is caused by rainwater leaks or condensation, it encourages biodeterioration of susceptible materials, wood decay being a particularly serious problem in some circumstances, as described in section 6.4, and perhaps requiring more expensive attention than repair of the felt roof covering.

It will be appreciated from these comments that, if a roof is designed to avoid the danger of interstitial condensation, a felt covering with a limited life is acceptable, provided that the limited life is recognised and the covering repaired or replaced at appropriate intervals. All building materials deteriorate, but they are only defective if they deteriorate more rapidly than expected. Natural stone or brick masonry is expected to deteriorate very slowly indeed, and premature deterioration is an indication of a defect, such as the use of inadequately durable stone or brick in relation to the conditions of exposure, or unsuitable mortar, or some abnormal effect, for example the development of sulphate attack in the mortar as described in section 7.3. In contrast, paint coatings on wood joinery are recognised as having only limited durability, and it is accepted that they require frequent maintenance.

Investigations into failures in connection with claims or disputes must obviously establish the causes of failure, and whether they can be attributed to normal deterioration or defects in construction or maintenance. Such investigations are not, of course, limited to site inspections but also involve laboratory examination and analysis of samples, as well as assessment of the situation in relation to construction or maintenance specifications, or normal good practice, usually represented by British Standard specifications and codes of practice. It is obviously essential that an investigation is thorough if the resultant report is to be authoritative, and there can usually be no excuse for failing to make a thorough investigation. This obviously involves a determined approach to the investigation, but also the use of appropriate equipment. In fact, equipment does not usually need to be very sophisticated, the most important requirements being suitable clothing, both weatherproof for external inspections in inclement conditions and overalls for internal inspections in dirty areas. Pencils are preferred for writing notes; pens do not write on damp notebooks! Torches are more efficient than lead lights, but they must be adequate and spare batteries must be available. Hand mirrors can be helpful for viewing inaccessible areas; they can usually be borrowed from mother's spare handbag! Borescopes can be very useful, although the more expensive elaborate models can rarely be justified; the simple Checkscope (Keymed Industrial Limited) is reasonably robust and particularly suitable. A moisture meter is essential but does not need to be particularly sophisticated; most of the simple probe instruments such as the Protimeter range will accurately measure wood moisture contents and will give relative measurements on other materials, enabling areas of high and low moisture content to be located. Temperature measurement facilities on the same instrument can be useful but are not essential. Convenience is perhaps the most important requirement; it may be found that a simple meter reading instrument, such as a Protimeter Minor, may be more convenient than a much more bulky meter or digital instrument; lights that illuminate at certain moisture contents are not sufficiently accurate or sensitive, although they may be useful in dark areas. Instruments with probes on leads require two hands in normal use and are not easy to use when working from ladders or in confined spaces.

Although inspections will often involve minimal equipment, such as a torch, moisture meter and Stanley knife, a complete set of house-breaking tools must be available as exposure works may be required, varying from lifting floor boards to removing bricks from walls. A minimal tool kit should include, in addition to the previously mentioned Stanley knife, a claw hammer, crowbar, wood chisel, small and large screwdrivers (both slot and Pozidrive), an Eclipse multi-purpose saw, lump hammer, cold chisels, bolster, an electric drill with various masonry and other bits and, finally, a sectional ladder. A cordless electric drill is particularly useful for taking plaster, masonry or wood samples for analysis, although a mains operated drill with a hammer facility is necessary for some operations, particularly drilling in dense concrete. If a mains drill can be used, high-speed wood bits and large diameter masonry bits can be used to provide access holes for borescope investigations. A slow-speed cordless drill is particularly useful for sampling wood for preservative analysis using a Forstner bit, which makes a flat bottom hole and thus samples to a required depth; tungsten-tipped drills of about 8 mm diameter can be used for similar sampling of masonry materials, such as for salt or mix analysis. Sharp tools must be protected; chisels can be purchased with an end guard in a plastic case, and Stanley knives are available with retractable blades. When working with sharp tools it is essential to carry also a first aid kit and to know how to use it, particularly when working alone.

Some observations are best recorded photographically, if possible. A camera fitted with a zoom lens, with a range of about 35 to 70 mm focal length with a macro facility will cover most requirements. A flash is essential; it is important to appreciate that flash settings must be adjusted to take account of the focal length of the lens. Although the use of colour negative film is attractive with the rapid printing facilities that are now available, printing involves the use of correction filters and colour balance cannot be assured. Properly exposed colour slide reversal film does not suffer from this disadvantage, and slides can be examined in detail on a screen if it is later necessary to check for features which may have been overlooked during the inspection; reversal prints can be obtained from colour slides which are similar in quality to prints from negative film. Colour prints pasted into reports can now be copied in accurate colour, greatly improving report presentation, and digital cameras now make it possible for all colour illustrations to be prepared in the office if a suitable computer facility is available.

There is no real difficulty in transporting a comprehensive range of equipment in a car, but difficulties arise when travelling by public transport, particularly by air where hand luggage is normally subject to a 5 kg weight limit and even checked luggage may be limited to 20 kg. The essential requirements are a reliable torch, a small moisture meter such as a Protimeter Minor, a Stanley knife with a retractable blade, plastic bags and envelopes for samples, and a folding umbrella which can be a great advantage for external inspections in inclement weather! Borescopes are often supplied by their manufacturers in boxes like small attache cases, which also accommodate the battery charger and spare bulbs; if it is necessary to limit weight and bulk, the battery can be fully charged in advance so that the charger is unnecessary and the probe itself can be provided with an alternative packing, such as a cardboard tube fitted with foam wrapping such as pipe insulation.

There are many other tools and instruments that may be helpful at times.

Although electrical conductivity moisture meters, such as the Protimeter range, can give very accurate measurements of moisture content in wood, their use on plaster and masonry can give very unpredictable results and it is often suggested that other methods of measurement are more suitable. The most accurate method for determining moisture content in any material is to weigh a sample before and after drying; as moisture must not be lost between sampling and initial weighing, sealed jars or bags are required. Alternatively, a carbide test method can be used, such as the Speedy moisture tester; a sample is mixed with carbide and the pressure of the generated acetylene gas gives an indication of the moisture content of the sample. In fact, since moisture content is not necessarily related to the 'dampness' of a material (section 4.2), accurate measurement of moisture content is not necessarily helpful in diagnosis, except for checking for the presence of hygroscopic salts (section 4.10). Most experienced investigators will find that they rarely use a Speedy or similar carbide moisture tester on site, and that intelligent use of an electrical conductivity moisture meter is sufficient, particularly for the accurate checking of wood components. This is usually sufficient to indicate the condition of the structure as a whole, as the moisture content of wood will directly indicate the 'dampness' of plaster and masonry with which it is in contact and, if wood is not in contact with other damp materials, the moisture content will depend directly on the relative humidity of the surrounding atmosphere. Electrical conductivity moisture meters suffer from the disadvantage that their probes cause holes in painted surfaces, but instruments without probes normally rely on capacitance to determine moisture content; results can be very inconsistent as instruments of this type are particularly subject to surface effects. One advantage of a needle probe moisture meter is the ability to check floor boards and screeds through carpet coverings or joints in tiles without causing damage.

1.4 Reports

An expert report on defects and deterioration must present an impartial assessment. The report cover page should show the identity of the property and the purpose of the investigation, as well as the identity of the instructing client. The report itself should be divided into several sections, basically a general introduction with a description of the situation, an account of the investigation, including any inspections and laboratory tests, a discussion and finally the conclusions; a further section on recommendations for remedial works may be appropriate in some circumstances.

The introduction may be very extensive, describing the background to the investigation, including appropriate references to the documentation, such as the contract and the specification. However, where the investigation forms part of a more extensive investigation by another person, the introduction can often be quite brief, referring simply to another report in which the information is already available. For example, if a structural engineer or building scientist is instructed to investigate a particular matter that has arisen in connection with a more general investigation carried out by an architect or building surveyor, the specialist report may take the form, in effect, of a supplement to the general report, although it is usually best to combine such reports if possible to form a single joint report and avoid unnecessary duplication. Extensive introductions are sometimes criticised for repeating information that is well known or available elsewhere, but if a report is expected to stand on its own, without any formal linking with any other document, the introduction must fully explain the circumstances in which the need for investigation arises, and it must be comprehensive rather than concise.

The second section of the report should be concerned with a factual account of the investigation, comprising site inspections and associated laboratory tests. It is very easy for the author of a report to omit obvious essential details, such as the dates of inspections, the persons present or the orientation of the property. If it is helpful, sub-headings should be provided to draw attention to different parts of the investigation or different parts of the property. The following section of the report should consist of the discussion. It will necessarily repeat in summary the results of the investigations in the previous section, but will then place the information in context, in relation, for example, to the specification, relevant Building Regulations, and normal good practice, usually as represented by British Standard specifications and codes of practice. However, in many cases an extensive discussion will not be justified and this section can be more sensibly included with the conclusions. Alternatively, if the investigation section includes a number of sub-headings relating to different parts of the investigation or different parts of the property, it may be more sensible to combine the investigation and discussion sections in order to group the substance of the report more conveniently.

It is recommended that the main sections of the report should not include extensive quotations from specifications or other documents, schedules, lengthy calculations or numerous diagrams; these are best presented as appendices, so that they do not interrupt the flow of the report which should be readable and readily understandable without the use of excessive scientific jargon.

Where investigations have been carried out in support of litigation or arbitration proceedings, it is sensible to use a word processor so that amendments can be incorporated as necessary without complete retyping. Experts often prepare an initial report, which leads to additional investigations perhaps by other persons, and progressive revision of the report. In some cases substantial corrections become necessary when documents are disclosed or following meetings of experts, and the final report for exchange between the parties is often completely different from the initial investigation report.

2 Structural problems

2.1 Introduction

The main function of a building is to protect the occupants and contents from the weather, principally rain, wind and extremes of temperature. It is most important to provide a simple intact envelope which will achieve all of these functions; features such as windows, flues, damp-proof courses and thermal insulation are only additional sophistications. Obviously a building must be structurally sound in order to survive, and intermediate suspended floors must be capable of resisting any normal imposed loads.

This book is not concerned with defects due to inadequate design or construction for required structural loadings, but structural failure can still occur, even with correct design and construction, if the materials used are unsuitable or subsequently deteriorate.

2.2 Structural failures

Structural failures are the result of over-stressing, that is, the imposition of loads in excess of the capacity of the structural components. Collapse is, of course, the ultimate and most serious result, but over-stressing is also evident at earlier stages through the development of distortion and fractures. If a structure is correctly designed and constructed in accordance with the design, over-stressing indicates some other inadequacy, such as the use of an unsuitable material. For example, in wood components it is necessary to make sure that the wood is a suitable species and grade to provide the required strength, as explained in section 6.2. However, even suitable materials may prove to be inadequate in some circumstances or because they have been altered in some way. For example, the design strength of wood assumes a normal 'dry' moisture content of about 12–15%, but strength is dramatically reduced if the wood has a significantly higher moisture content; the strength is halved if the moisture is at the fibre saturation point of about 28% or above. The elasticity of wood is also altered and creep or progressive distortion can occur under load. The strength of wood also declines with age. In fact, none of these factors are commonly involved in over-stressing, which usually results from damage of some sort, such as fungal or insect deterioration, as described in sections 6.4 and 6.5, or the equally serious deliberate notching of joists and beams, often known as plumber's or electrician's 'rot'.

An experienced surveyor will often check the apparent stability of suspended floors by feeling the deflection when subjecting them to heel bumping, a sluggish response indicating over-stressing. This will suggest the necessity for further investigation, perhaps by lifting the floor boards to expose the supporting joists and beams. It may be discovered that the dimensions of the supporting components are inadequate for the imposed loads, or insect borer or fungal decay damage may be discovered. However, the most common cause of weakening is notching of joists and beams by thoughtless plumbers and electricians. Holes drilled at mid-depth will have no significant effect on strength; while such holes can be used by electricians for cable runs, the situation is more difficult for plumbers who prefer, for obvious convenience, to run their pipes in notches cut into the tops of joists and beams. In domestic properties it is common to see notches 25 mm deep in 150 mm deep joists; the effect is to reduce the depth of the joists to only about 125 mm and thus seriously weaken the joists. These depend upon the square of the depth for their strength; a 125 mm deep joist has only about 70% of the strength of a 150 mm joist, so that typical plumber's 'rot' causes a loss of strength of about 30%. Excessively wide notching or thoughtless attention to pipe runs can also result in inadequate support for floor boards.

Alterations can cause similar problems in buildings on a rather larger scale. In most buildings the accommodation comprises a series of boxes in which cross walls provide bracing or racking resistance. In older houses ground-floor partition walls are often removed to combine the original front sitting room and rear dining room into a single large living room, and in commercial buildings, particularly shops, alterations are often concerned with providing large unobstructed areas by replacing partitions with isolated posts and large beams. In an extreme case, which actually occurs very frequently in main shopping areas, a large shop may consist only of two flank walls, with the entire space between them spanned by large beams and perhaps intermediate posts, but with no bracing or racking resistance. The buildings do not usually collapse because they are supported on either side by neighbouring buildings, but an entire street of buildings might collapse with the removal of the supporting building at one end. The situation is usually worst in shopping areas where old buildings have been converted, particularly where planning restrictions have resulted in the retention of structures unsuitable for modern usage.

Structural stability problems do not result only from the removal of bracing partitions. Attempts to utilise roof spaces often result in cutting new doorways through truss tie beams; the feet of the truss rafters then spread, perhaps distorting the heads of their supporting walls. Struts and wind braces are often removed from roofs to allow unobstructed accommodation, and purlins are cut to allow for new dormers; all these alterations weakening the structure, although perhaps becoming apparent only under extreme wind or snow loads. Similar events may also cause failures attributable to inadequate original construction, perhaps many years after construction is completed. Perhaps the most common failure of this type is wall-head distortion caused by spreading rafter feet which are not properly tied. The ceiling joists normally tie the common rafters, but in some forms of construction, such as with changes in levels between one side of a building and the other, different tie arrangements are required; if they are not properly provided, spread generally occurs. Old roofs often do not show distortion until, for example, the original slate covering is replaced with the much heavier concrete tiles that are now favoured because of their lower cost and easier availability.

Masonry can severely deteriorate if the brick or natural stone from which it is constructed is insufficiently durable. Deterioration may take the form of progressive erosion or spalling, perhaps involving the loss of a surface layer as much as 10 mm thick in each spall. Damage of this type is obviously most severe in very exposed components, such as copings, cornices and pinnacles, and the resulting structural damage is most severe in small section components, again including pinnacles, and also relatively thin features, such as window mullions and transoms. Masonry deterioration is associated with moisture and is described in sections 4.3, 4.4, 7.2 and 7.3.

Masonry deterioration is not restricted to brick or natural stone components, but can also occur in mortar joints. Mortar is effectively synthetic stone which can suffer deterioration in the same way as natural stone; in addition, mortar usually contains cement, and severe expansion and loss of strength can occur if soluble sulphates react with the tricalcium aluminate component in ordinary Portland cement, a form of deterioration known as sulphate attack and described in more detail in section 4.4.

Sulphate attack is not confined to mortar used in joints, renders or screeds, but can also occur in concrete similarly affected by soluble sulphates, as explained in more detail in section 8.6. Chloride is also a problem in reinforced concrete as it encourages the rusting of steel reinforcement which swells and fractures the concrete, as explained in section 4.5; reinforcement corrosion is inhibited by the alkalinity of concrete, but develops as the alkalinity is neutralised by carbonation, that is the absorption of carbon dioxide from the atmosphere, an effect that may occur excessively rapidly if concrete is physically porous or the cover of the concrete over the reinforcement is inadequate. The presence of chloride will exaggerate the rate of corrosion; whilst chloride may be derived from natural sources such as sea spray, most problems are associated with the excessive use in the past of calcium chloride as a concrete curing accelerator. Although the residual effects of calcium chloride misuse probably accounts for most failures in structural concrete, failures due to the use of high alumina cement certainly generated much greater publicity; in humid atmospheres, such as roofs of swimming pools, concrete made with high alumina cement can suffer serious loss of strength, as explained in more detail in sections 4.4 and 8.2.

Metal corrosion can take many forms, as explained in section 4.5, but it usually only results in structural failure when critical fixings corrode, or when the corrosion products cause destructive expansion damage. This is the form of failure that has been previously described for concrete reinforcement, and is certainly the most common form of structural damage from metal corrosion in buildings, although corrosion damage of ties in cavity walls can also result in weakening; this is often the reason why gables collapse in storms.

It will be apparent from these comments that structural failures are not themselves defects, but result from defects or deterioration. Failures may be apparent but their causes may be much less obvious.

2.3 Structural movement

Although a building structure is normally designed to be rigid and inflexible, a considerable amount of movement actually occurs. The main causes of movement are simple expansion and contraction due to changes in temperature or moisture content, as described in sections 3.5 and 4.2 respectively. Such movement is not usually too serious if it affects the entire structure uniformly, but problems can arise through differential movement between parts of the structure differently affected by temperature or moisture content, or parts constructed in different materials which respond in different ways. In a very long structure, shrinkage may cause tensions which result in fractures, and both design and construction should allow for this movement by the provision of vertical movement joints at regular intervals. Internal problems in buildings are usually associated with differential movement due to contrasting properties of adjacent materials. A concrete building will generally shrink as it cures, this shrinkage being fully developed only after perhaps a year or two; installed materials which are too rigid to absorb this shrinkage must be provided with movement joints, the most obvious examples being granite, marble or ceramic tiles on walls and floors. With wood strip or block floors, the wood should be dried or seasoned to the average moisture content that it will achieve in service before it is installed. If a wood floor is exposed to excessive humidity after it is laid it will swell, and the floor may become seriously distorted, subsequently drying in service causing gaps to form if movement joints are omitted.

Moisture content variations are the usual cause of movement in structural materials, such as concrete, masonry and wood, but only thermal movement is significant with metals. Steel beams and lintels are not often the cause of differential movement problems in masonry, but metals subject to extremes of temperature often cause problems. The main problems in normal service involve hot water and heating pipes, but in a fire initial heating often causes considerable expansion and severe damage to a structure, usually sufficient to prevent repair even if the fire is extinguished at an early stage; if the fire proceeds further, steel beams tend to soften and collapse under load, pulling in and destroying the walls previously fractured by their expansion. It would seem obvious that, as wood is combustible, the use of steel in buildings would be safer in the event of fire, but wood beams do not suffer from these destructive expansion and loss of strength problems. Wood is actually very durable in a fire because its thermal insulation properties ensure a slow fire penetration rate and continuing structural integrity. The combustibility of large wood components, such as beams, does not usually contribute significantly to a fire; it is the furnishings and other components in a building that represent the main fire risk and contribution to fire damage.

Movement with moisture content and temperature changes is a normal problem which must be allowed for in design and construction, but there are other forms of movement which arise through defects or deterioration. The expansion of mortar and concrete due to sulphate attack, and the rupture damage to concrete due to the expansive corrosion of steel reinforcement have both been mentioned in the previous section 2.2 and are described more fully in sections 4.4, 4.5, 8.5 and 8.6.

Settlement and subsidence are significant causes of movement damage in buildings. Settlement is a natural process involving compaction of the supporting soil under the load of a building. Settlement during construction always occurs to a certain extent and is not usually too troublesome, provided that it is uniform over the entire supported area of the building. The constant applied load may, however, result in creep and some further progressive settlement over the early life of a building which may cause difficulties if the building is, for example, an extension linked to an existing stable building. If the support varies over the base of the building, differential settlement may occur resulting in fractures in the structure, and it will be necessary to consider whether these fractures threaten the stability of the building. Obviously the load-bearing properties of the supporting ground must be considered when designing and constructing a building, and if settlement problems are likely to occur it is necessary to take special precautions, such as the construction of foundations that are deeper than normal or even piling, or alternatively the provision of a rigid raft structure. The greatest settlement problems usually arise with peaty soils.

Subsidence is settlement resulting from factors other than simple compression of the supporting ground. The most dramatic subsidence results from mining and other tunnelling operations beneath a building, as well as erosion of the supporting ground due to water movement, due either to natural water flow or leaking water mains or drains, the latter often causing local subsidence of, for example, the corner of a building. However, subsidence can also occur through the building itself interfering with the ground conditions, particularly on shrinkable clay where the building and its drainage system will reduce rain penetration into the soil which will dry out and shrink as a result. Shrinkable clay can also cause seasonal movement, particularly in abnormally dry summers such as 1976 and 1989; where such clay occurs it is essential to deepen the foundations to provide support beneath the clay layer, or at least sufficiently deeply within the clay to avoid these effects that are most serious near the surface. Shrinkable clay is also a particular problem in association with trees.

Foundation problems, including settlement, subsidence and tree root damage, are discussed more fully in Chapter 10.

Thermal problems

3

3.1 Introduction

One of the main functions of a building is to isolate the accommodation from the external temperature conditions. The most important structural factor in achieving this function is the actual envelope enclosing the accommodation, even if the envelope has no thermal insulation value, as there is always resistance to the transfer of heat from the outside air to the envelope structure and from it to the inside air, or vice versa. This resistance to heat transfer is known as surface resistance and applies, whatever the thermal properties of the envelope; the actual level of resistance varies depending on the type of surface, so that a polished reflective surface will resist heat transfer and will have high surface resistance compared to a rough surface. Further resistance to heat transfer can be achieved by improving the thermal insulation of the structural envelope, so that resistance to structural heat loss will depend on the surface resistance plus the structural resistance, as explained in more detail in section 3.2. Heat loss does not, however, depend only on structural heat loss, but also on ventilation heat loss through both deliberate and accidental ventilation, as explained more fully in section 3.3.

Thermal problems in buildings in the British Isles are usually considered to be associated only with winter conditions and the need to improve insulation to reduce heat loss and therefore to reduce energy cost and to improve comfort. Whilst these aims are obviously sensible and advantageous, energy crises have prompted the Government to extend the Building Regulations beyond their original health and safety scope to include energy conservation. Obviously, precisely similar problems arise in reverse in tropical countries, where insulation is required to improve the efficiency of air conditioning cooling systems. Thermal factors influence many aspects of a building: obviously ventilation, but also the development of condensation as mist in the accommodation air, dew on the accommodation surfaces and interstitial condensation within the structural elements; these will, in turn, affect the thermal insulation efficiency of the envelope and perhaps introduce a risk of fungal decay in wood and other susceptible materials. In addition, temperature variations in buildings cause thermal movement, as discussed more fully in sections 2.3 and 3.5.

In recent years the levels of structural heat loss permitted under the Building Regulations have been steadily reduced, mainly in an attempt to reduce national energy consumption for space heating, although economies have not been fully realised as occupants have taken advantage of improved insulation to maintain higher temperature levels and greater degrees of comfort. In recent years the Building Regulations have allowed several different methods for the assessment of heat loss from a building, as explained in Approval Document L, but these options do not affect the ways in which heat is lost and the potential defects considered in this chapter.

It is relatively easy to design new building elements to avoid excessive structural heat loss, and even to carry out remedial improvements, such as the addition of loft insulation, but it is much more difficult to reduce ventilation heat loss, other than by draught-proofing. There is generally excessive ventilation and thus excessive heat loss in buildings provided with flues, while inadequate ventilation in flueless buildings can lead to stuffiness and condensation. Ventilation heat loss can be minimised, but sufficient ventilation is essential for safety and comfort, as discussed in more detail in sections 15.2 and 18.2.

Even if thermal insulation is perfect and structural heat loss is minimised, some ventilation heat loss will still occur because ventilation is essential. It is easy to reduce heat loss substantially from a structure that is poorly insulated, but it becomes progressively more difficult to achieve further reductions, and the cost must be balanced against the financial savings that are likely to be achieved. These economic factors, considered in more detail in section 3.4, are often ignored, particularly by firms who are keen to sell double glazing and other remedial systems, and who usually publicise only the percentage reduction in heat loss through an individual element that their system will achieve without relating the saving to the total heat loss from the building. For example, efficient double glazing will reduce heat loss through a window by about 50% but, as heat loss through single glazed windows represents only about 30% of the total heat loss from an average modern house, and perhaps as little as 18% from an older house, the actual savings are only 15 or 9% respectively.

3.2 Structural heat loss

The rate of heat loss from a building element is usually specified in terms of the thermal transmittance or U value, defined as the rate of heat transfer (watts) through unit area (one square metre) of the building element for unit temperature difference (one degree kelvin or centigrade), so that the units for U value are $W/m^2\,^\circ K$ or $W/m^2\,^\circ C$. (Kelvin and Centigrade degrees are identical, but the Kelvin notation, in which the freezing point for water is $273\,^\circ K$ and boiling point is $373\,^\circ K$, is used in the current Building Regulations, presumably to confuse architects, engineers and contractors who have only recently absorbed the transition from Fahrenheit to Centigrade degrees!) For convenience the Centigrade notation will be used throughout the text as it is more widely understood.

The advantage of defining the thermal properties of a building element in terms of thermal transmittance or U value is that the energy loss through the element can be calculated by multiplying the U value by the area and the temperature gradient. For example, if an exposed wall has a U value of 0.6 and an area of $25\,m^2$, the total energy loss through the wall will be $15\,W$ for a difference in temperature of one Centigrade degree, so that if the average temperature difference over the year between the interior accommodation and the exterior air is $10\,^\circ C$, the rate of heat loss is $150\,W$, representing a total energy loss over the year of $1314\,kW/h$. It is obviously in the national interest to limit energy consumption for heating buildings, and the current Building Regulations therefore limit U values; the requirements for dwellings are summarised in Table 3.1. As the purpose of these requirements is energy conservation, it is essential in an element of mixed construction, such as a wall with windows, to achieve the overall requirement, even if individual parts cannot comply for some reason. If the proportion of windows is excessive, the U value of the windows and/or the walls must be reduced in compensation, as explained in the notes to Table 3.1. For example, a house with $100\,m^2$ of wall

Table 3.1 Maximum thermal transmittance (*U*) values for elements of dwellings

Building element	Maximum (U) value (W/m² °C)
Exposed walls and floors	0.6
Roof	0.35
Windows and roof-lights	5.7

Notes
1. These requirements are based on the Building Regulations 1985 for England and Wales.
2. Degrees Kelvin (°K) are used in the Building Regulations but they are identical to degrees Centigrade (°C).
3. The area of the windows and roof-lights shall not in effect exceed 12% of the external walls; if their area is greater, the maximum *U* values of one or more of the elements must be reduced in compensation.

meeting the critical requirements will have $12\,m^2$ of windows with a U value of $5.7\,W/m^2°C$ and heat loss of $68.4\,W/°C$, and $88\,m^2$ of wall with a U value of $0.6\,W/m^2°C$ and heat loss of $52.8\,W/°C$, giving a total heat loss of $121.2\,W/°C$, which must not be exceeded. If the window area is increased to $20\,m^2$ with single-glazed windows with a U value of $5.7\,W/m^2°C$ and a heat loss of $114.0\,W/°C$, the wall U value must be reduced to below $0.09\,W/m^2°C$ to ensure that total heat loss still does not exceed $121.2\,W/°C$. In fact, a U value of $0.09\,W/m^2°C$ cannot be achieved with conventional construction, so that single-glazed windows cannot be used if the window area is 20% of the wall area. The solution to this dilemma is to use double glazing. A $20\,mm$ air gap and a well insulated wood, uPVC (unplasticised polyvinyl chloride) or thermal break aluminium frame will give a U value of $2.5\,W/m^2°C$ and a heat loss for $20\,m^2$ of $50\,W/°C$, the wall requirement then being a heat loss not exceeding $71.2\,W/°C$; this represents for $80\,m^2$ a U value not exceeding 0.89, which is easily achievable.

Although thermal transmittance or U values are the most convenient way in which to express heat loss from a building, they cannot be directly determined from the properties of the various components or layers of a structural element. The converse of thermal transmittance or U value is thermal resistance or R value, these two values being reciprocals; it is R value that can be calculated most easily, as the thermal resistance of each succeeding layer can be added to give the total R value of the element. The simplest situation is represented by a single structural layer, such as single glazing, but in addition to the resistance of the material itself, there are also external and internal surface resistances which represent the resistance to transfer of heat from air to solid material. Typical surface resistances are shown in Table 3.2, from which it can be seen that surface resistance depends on the surface emissivity, or the efficiency of the surface in transferring heat, high emissivity representing normal surfaces and low emissivity representing reflective surfaces, such as polished aluminium window frames and aluminium foil backing on plaster board. Surface resistance also depends on the direction of heat flow; this is horizontal through walls and windows, upward through ceilings and roofs (except in summer conditions or warm climates when heat flow is downwards), and downwards through ground floors. The resistance of external surfaces also depends on the degree of exposure, normal exposure being adopted except in special cases. As far as the example of single glazing is concerned, the internal surface resistance involves high emissivity and horizontal heat flow, and thus a surface resistance of $0.123\,m^2°C/W$. The external surface resistance for high emissivity and normal exposure is $0.055\,m^2°C/W$.

The thermal resistance of a layer of material depends on thickness divided by

Table 3.2 Surface resistances (m² °C/W)

Internal surface resistance R_{Si}

Building element	Surface emissivity	Heat flow	Surface resistance
Walls	high	horizontal	0.123
	low	horizontal	0.304
Roofs, ceilings, floors	high	upward	0.106
	low	upward	0.218
	high	downward	0.150
	low	downward	0.562

External surface resistance R_{So}

Building element	Surface emissivity	Surface resistance		
		sheltered	normal	exposed
Walls, windows, doors	high	0.08	0.055	0.03
	low	0.11	0.067	0.03
Roofs, floors over ventilated areas	high	0.07	0.045	0.02
	low	0.09	0.053	0.02

Notes
1. Values from Building Research Digest 108 (IHVE Guide, book A).
2. Emissivity is high for all normal building materials including glass, but unpainted metal and other reflective surfaces have low emissivity.
3. External surface resistance depends upon exposure:
   Sheltered: up to 3rd floor in city centres.
   Normal: most sites,
              4th to 8th floors in city centres.
   Exposed: coastal or hill sites,
             9th floor and above in city centres,
             5th floor and above elsewhere.

thermal conductivity (k); Table 3.3 lists typical k values for building materials. Continuing the example of single glazing, the glass is perhaps 3 mm thick (0.003 m) and, as the k value from Table 3.3 is 1.05, the resistance of the glass itself is about 0.003.

The calculations for single glazing can be conveniently summarised as follows:

$$R = R_{So} + R_G + R_{Si}$$

Thermal calculations for single glazing

External surface resistance (high, normal)	$R_{So} = 0.055$
Glass resistance (3 mm, $k = 1.05$)	$R_G = 0.003/1.05 = 0.003$
Internal surface resistance (high)	$R_{Si} = 0.123$
Total thermal resistance	$R = 0.181$
Thermal transmittance (reciprocal)	$U = 5.53$

It is apparent from this calculation that the surface resistances are far more significant than the insulation properties of the glass, indicating that it is the envelope around accommodation that provides the most important isolation from the surroundings. The reciprocal of the thermal resistance or R value is the thermal transmittance or U value, which is 5.53 in this case for single glazing alone. In fact,

Table 3.3 Typical thermal conductivity (*k*) values

	Material	Moisture content	Bulk density	Thermal conductivity
Asbestos	Cement sheet	5	1600	0.4
	Insulation board	5	750	0.12
Asphalt	Roofing		1920	0.58
Brick	(See Fig. 3.1)			
	Common, dry	0	1760	0.81
	Conditioned at 17.8°C			
	and 65% RH	6	1870	1.21
	Wet	16	2034	1.67
Building				
board	Asbestos insulator	2	720–900	0.11–0.21
	Fibre board		280–420	0.05–0.08
	Hardboard, medium		560	0.08
	Plasterboard, gypsum		1120	0.16
	Woodchip board		350–1360	0.07–0.21
	Woodwool slab	5	400–800	0.08–0.13
Carpeting	Wilton type			0.058
	Wool felt underlay		160	0.045
	Cellular rubber underlay		270	0.065
			400	0.10
Concrete	(see Fig. 3.1)			
	Gravel 1:2:4		2240–2480	1.4
	No fines, gravel 1:10		1840	0.94
	Clinker aggregate	4	1680	0.40
	Expanded clay			
	aggregate	5	800–1280	0.29–0.48
	Pumice aggregate	4.6	770	0.19
	Vermiculite aggregate		400–880	0.11–0.26
	Cellular		320–1600	0.08–0.65
Cork	Granulated, raw	7	115	0.046
	Slab, raw	7	160	0.050
	Slab, baked	3–5	130	0.040
Felt	Undercarpet felt		120	0.045
	Asbestos felt		144	0.078
	Roofing felt		960–1120	0.19–0.20
Glass	Sheet, window		2500	1.05
	Wool, lightweight mat		25	0.04
Insulation				
materials	Aerated concrete block			
	or slab	3	600	0.19
	Aerated concrete block			
	or slab	3	400	0.14
	Glass fibre quilt or mat		50	0.033
	Granulated expanded			
	polystyrene		12–16	0.038
	Pelletized glass fibre		40	0.04
	Mineral wool		50	0.045
	Expanded vermiculite		100	0.065
	Polyurethane foam		30	0.026
Metals	Aluminium alloy, typical		2800	160
	Brass		8400	130
	Copper 99.9%		8900	200
	Iron, cast		7000	40
	Lead		11340	35
	Steel, mild		7850	47

Table 3.3 continued

	Material	Moisture content	Bulk density	Thermal conductivity
	Steel, high alloy		8000	15
	Zinc 99.9%		7130	113
Plaster	(see Fig. 3.1)			
	Gypsum		1120–1280	0.38–0.46
	Perlite aggregate		400–610	0.079–0.19
	Vermiculite aggregate		480–960	0.14–0.30
	Sand cement		1570	0.53
Plastics, cellular	Polystyrene, expanded board		15	0.037
	Polyurethane foam		30	0.026
	Polyvinyl chloride, rigid foam		25–80	0.035–0.041
	Urea formaldehyde foam		8–30	0.032–0.038
	PVC floor covering		0.40	2.5
Plastics, solid sheet	Acrylic resin		1440	0.20
	Nylon		1100	0.30
	Polycarbonate		1150	0.23
	Polyethylene, low density		920	0.35
	Polyethylene, high density		960	0.50
	Polypropylene		915	0.24
	Polystyrene		10.50	0.17
	PTFE		2200	0.24
	PVC rigid		1350	0.16
Roofing felt				
Stone	(see Fig. 3.1)			
	Granite		2650	2.9
	Limestone		2180	1.5
	Marble		2700	2.5
	Sandstone		2000	1.3
	Slate		2700	1.9
Tiles	Burnt clay		1900	0.85
	Concrete		2100	1.10
	Cork		530	0.085
	PVC Asbestos		2000	0.85
	Rubber		1600–1800	0.30–0.50
Vermiculite	Loose granules		100	0.065
	Plastering		480–960	0.144–0.303
Wood	*Across grain*			
	Beech	15	700	0.165
	Mahogany	10	700	0.155
	Oak	14	770	0.160
	Teak	10	700	0.170
	Pine	12	610	0.125
	Spruce	12	420	0.105
	Along grain			
	Oak	14	770	0.290
	Pine	12	610	0.215

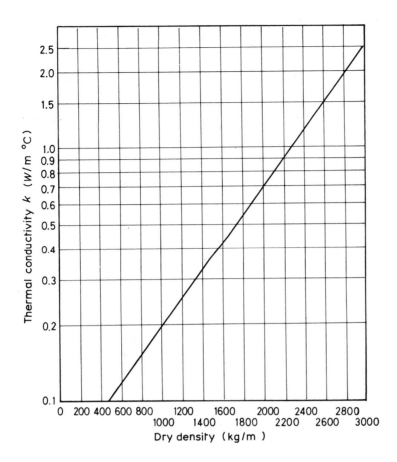

Fig. 3.1 Variation of thermal conductivity k with density. The graph shows the thermal conductivity for typical dry brickwork, masonry, concrete and plaster, but thermal conductivity increases with moisture content and a correction must be applied by multiplying by the appropriate factor.

Exposure conditions and materials	Moisture content (percentage by volume)	Correction factor
Protected brickwork, granite, sandstone	1	1.3
Protected concrete, limestone, plaster	3	1.6
Exposed brickwork, concrete, masonry	5	1.75
Severe exposure to driving rain, porous masonry	10	2.1

Table 3.4 Thermal transmittance (U) values for windows

Construction	Frame proportion	U value (W/m2 °C)		
		sheltered	normal	exposed
Metal, single-glazed	20%	5.0	5.6	6.7
Wood or uPVC, single-glazed	30%	3.8	4.3	4.9
Metal, double-glazed (6 mm)	20%	3.6	3.8	4.4
Metal thermal break, double-glazed (20 mm)	20%	3.0	3.2	3.5
Wood or uPVC, double-glazed (20 mm)	30%	2.3	2.5	2.7

windows are multi-component elements consisting of frames as well as glazing. The U value for the frame, which usually occupies 20 to 30% of the window area, depends upon its construction; the U value for a normal uninsulated aluminium or painted steel frame is about 5.5, the same as for the glass. Examples of U values for complete frames are given in Table 3.4. An interesting point is the relatively high U values for proprietary metal frames with factory-sealed double glazing with a relatively small air gap, but U values are greatly reduced, indicating improved thermal insulation, with wider air gaps and more efficient wood, uPVC or thermal break aluminium frames. The importance of a sufficient air gap is illustrated in Figure 3.2.

In more complex forms of construction exactly the same principles are involved in which the total thermal resistance or R value is determined by adding the resistances for the surfaces and each successive layer of the structural element. In some cases cavities occur, their thermal resistances depending on the emissivities of the surfaces on either side of the cavity, as well as the ventilation, as summarised in Table 3.5.

The thermal resistance or R value for a composite building element can be calculated using the general formula

$$R = R_{S_o} + R_{S_i} + R_C + R_1 + R_2 + \ldots$$

For example, the thermal resistance or R value, and its reciprocal the thermal transmittance or U value, for a typical external cavity wall can be calculated as follows.

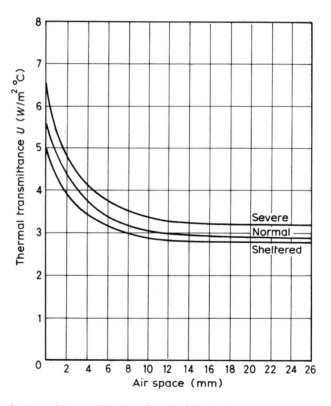

Fig. 3.2 Thermal transmittance U values for double glazing.

Table 3.5 Typical cavity resistances (m² °C/W)

Building element	Surface emissivity	Heat flow	Thermal resistance
Unventilated airspaces			
5 mm airspace	high		0.11
	low		0.18
20 mm minimum airspace	high	horizontal/upwards	0.18
		downwards	0.21
	low	horizontal/upwards	0.35
		downwards	1.06
Ventilated airspaces, minimum 20 mm			
Airspace between asbestos-cement or dark painted metal cladding with unsealed joints over high emissivity surface such as brickwork			0.16
As above but over low emissivity surface such as foil-backed plasterboard			0.30
Loft space between flat ceiling and unsealed asbestos-cement or dark metal cladding pitched roof			0.14
As above but aluminium cladding instead of black metal or with low emissivity upper surface on ceiling such as foil-backed plasterboard			0.25
Loft space between flat ceiling and unsealed tiled pitched roof			0.11
Airspace between tiles and sarking felt on pitched roof			0.12
Airspace behind tiles on tile-hung wall			0.12

Notes
1. Values from Building Research Digest 108 (IHVE Guide, book A)
2. Emissivity is high for all normal building materials, but unpainted metal and other reflective surfaces have low emissivity.
3. The thermal resistance values include the surface resistances on either side of the cavity.

Thermal calculations for typical cavity wall

External surface resistance (wall, high, normal) $R_{So} = 0.055$

Internal surface resistance (wall, high) $R_{Si} = 0.123$

Outer brickwork (0.1 m, 1700 kg/m³, 5% moisture content) $R_1 = 0.1/(0.47 \times 1.75) = 0.122$

Cavity resistance (20 mm +, high, horizontal) $R_C = 0.180$

Inner blockwork (0.1 m, 600 kg/m³, 3% moisture content) $R_2 = 0.1/(0.13 \times 1.6) = 0.481$

Plaster (0.01 × m, 600 × kg/m³, 3% moisture content) $R_3 = 0.01/(0.13 \times 1.6) = 0.048$

Total thermal resistance $R = 1.009$

Thermal transmittance (reciprocal) $U = 0.991$

It can be seen from these calculations that the lightweight inner blockwork, with a density of 600 kg/m³, compared with about 1700 kg/m³ for normal brickwork or dense concrete blocks, provides the main contribution to thermal resistance and is thus most significant in minimising the thermal transmittance or U value. The use of a lightweight block inner skin and lightweight plaster was sufficient for many years to keep the U value below 1.0 and thus meet the requirements of the Building Regulations, but the 1985 Regulations required the U value to be less than 0.6. This can be achieved only by the use of cavity fill insulation, built in as sheets or installed by injection in standing walls. Cavity fill insulation, such as granulated expanded polystyrene and pelletised glass fibre, will increase the cavity resistance R_C from 0.180 to about 1.30, increasing total resistance R to 2.129, giving a thermal transmittance or U value of 0.47, well within the current limit of 0.6. Mineral wool is less effective but polyurethane foam is more effective. When the insulation is incorpo-rated in new buildings as thick sheets or bats, the sheets are not usually as thick as the cavity and are secured to the inner leaf, leaving a narrow cavity between the insulation and the outer leaf, slightly reducing the thermal insulation but keeping it within the necessary limits, and also largely avoiding the moisture penetration prob-lems that can arise when the cavity is fully filled with insulation.

Thermal resistance R and transmittance U values can be calculated for any other structural elements in the same way. For example, the following calculation is for a timber frame wall, finished internally with plasterboard on a polythene vapour barrier over glass fibre quilt insulation, with a cavity, then sheathing plywood covered externally with breather paper, and finally counter battens, battens and ver-tical tiling. In thermal calculations the polythene vapour barrier and the breather paper are ignored as their effect is insignificant, and the calculation is made through the panels which comprise most of the wall area, as the wood frames or studs have good insulating properties, comparable with the cavity and glass fibre quilt. The vertical tiling cavity resistance is taken from Table 3.5.

Thermal calculations for tile-clad timber frame wall

External surface resistance (wall, high, normal)	$R_{So} = 0.055$
Internal surface resistance (wall, high)	$R_{Si} = 0.123$
Vertical tiles (0.015 m, $k = 0.85$)	$R_1 = 0.015/0.85 = 0.018$
Tile cavity	$R_{C1} = 0.120$
Plywood (0.012 m, $k = 0.125$)	$R_2 = 0.012/0.125 = 0.096$
Frame cavity	$R_{C2} = 0.180$
Glass fibre insulation quilt (0.050 m, $k = 0.033$)	
	$R_3 = 0.050/0.033 = 1.515$
Plasterboard (0.010 m, $k = 0.16$)	$R_4 = 0.010/0.16 = 0.063$
Total thermal resistance	$R = 2.170$
Thermal transmittance (reciprocal)	$U = 0.461$

The exceptionally high thermal resistance of the glass fibre insulation quilt is very apparent from these calculations, but it is most apparent if the thermal resistances of the various layers of the wall are plotted graphically, as in Figure 3.3. The tem-perature gradient is in proportion to the thermal resistance so that the temperature loss across the building element can be plotted on this graph as a straight line, thus indicating the temperatures at the interfaces between various layers of the structural element. If these temperatures are then transferred to the interfaces on a diagram of a section through the wall and the points joined up to show the temperatures through the wall, the steep slopes indicate the best insulation which is, in this case, across the glass fibre insulation quilt. Diagrams of temperature drop across a build-ing element are thus useful as an indication of the thermal significance of the various components, but they are also useful in assessing the dangers of interstitial conden-sation, as discussed in greater detail in section 4.7.

In this particular example, illustrated in Figure 3.3, in which the internal temper-ature is 20 °C and the external temperature is 0 °C, it is clear that if the internal air has a humidity equivalent to a dew point of 11 °C, interstitial condensation will occur where the structure is at lower temperatures, that is from within the insula-tion to the exterior of the element. Precautions are therefore necessary to prevent this condensation from accumulating and causing fungal decay in the sheathing plywood and stud frames. The usual precautions are to provide an impermeable vapour barrier on the warm side of the main insulation to prevent diffusion of warm humid air from the accommodation into the building element, coupled with a per-

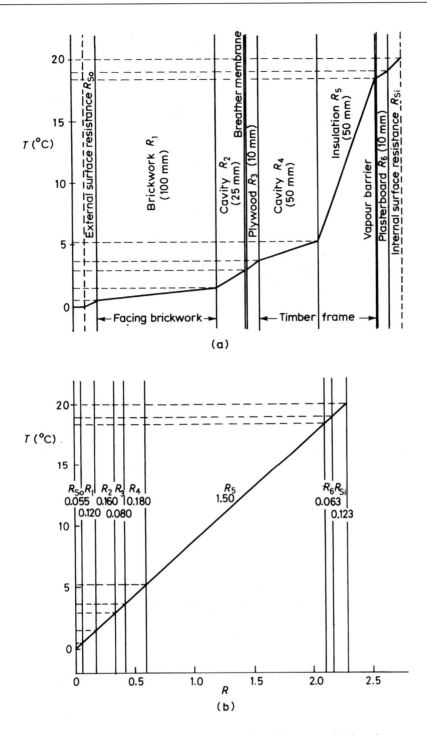

Fig. 3.3 (a) Section through brickwork-clad timber frame wall showing temperature gradients through each component. (b) Temperature plotted against thermal resistance to determine temperatures at component interfaces.

meable breather paper on the outside of the sheathing plywood which will allow any condensation to disperse to the exterior through the ventilated cavity beneath the vertical tiling. If the internal vapour barrier is omitted there is still a danger of condensation damage, as the rate of condensation formation will be too great in comparison with the rate at which it can be dispersed by diffusion through the sheathing plywood and breather paper; in such circumstances the danger can only be avoided by introducing some limited ventilation into the cavity between the glass fibre insulation quilt and the sheathing plywood. The importance of such studies on building elements can be appreciated from the fact that the vapour barriers are often damaged during construction, particularly by electricians. It is therefore safer, if possible, to avoid dependence upon a vapour barrier and to rely instead upon ventilation on the cold side of the insulation.

This is precisely the situation that exists in a conventional roof structure where the roof space is ventilated and condensation problems are never encountered. However, in order to meet modern requirements, sarking felt must be introduced beneath the slating or tiling battens and the thermal insulation properties must be improved. The ventilation of the roof space must be maintained in order to avoid interstitial condensation dangers, usually by providing ventilators in eaves soffits, but the function of this ventilation in terms of interstitial condensation control will only be properly maintained if the main insulation is provided on the ceiling so that the roof space is cold and freely ventilated.

A normal pitched tile roof with sarking felt and the ceiling beneath formed with foil-backed plasterboard, has a typical R value of about 0.66, or U value of about 1.5. The addition of 50 mm glass fibre quilt between the ceiling joists is sufficient to increase the R value to about 2.0, reducing the U value to about 0.5 which was sufficient for many years to meet the requirements in the Building Regulations. However, the 1985 Regulations reduced the U value limit to 0.35, equivalent to an R value of about 2.86. Assuming that glass fibre quilt has a k value of about 0.033, these requirements can be met by using glass fibre quilt more than about 72 mm thick; in fact, it is normal now to use 100 mm quilt. The critical thicknesses of other insulations, compared with 72 mm of glass fibre quilt, are about 82 mm for granulated expanded polystyrene, 88 mm for pelletised glass fibre, 99 mm for mineral wool, and 142 mm for vermiculite; obviously vermiculite and mineral wool are not really suitable to meet modern requirements in view of the excessive depths of material that are required.

The same principles apply to flat roofs, whether they are constructed from timber, steel or concrete. Their properties are calculated in precisely the same way as for the earlier cavity wall and timber frame wall examples, although with suitable corrections to the surface resistances, and diagrams similar to those in Figure 3.3 can be prepared to show the temperature gradients and the condensation dangers associated with dew point positions under appropriate conditions. However, the importance of designing to avoid interstitial condensation is not widely understood. In timber roofs it is possible to construct a normal 'cold' roof, that is a roof in which the main insulation is on the ceiling and there is a cold ventilated space above the insulation. The alternative is a 'warm' roof in which the main insulation is on the roof deck, immediately beneath the external roof covering. In a warm roof, interstitial condensation within the insulation is inevitable, unless warm humid air from the accommodation is prevented from diffusing into the insulation by provision of a reliable vapour barrier on the warm side of the insulation. It is virtually impossible to provide a sufficiently reliable vapour barrier at ceiling level, as there is always a danger that air can diffuse round the edges of the barrier and that electricians will damage the barrier when installing light fittings. A far more reliable arrangement is to provide the vapour barrier on the roof deck, but it would be apparent from plotting temperature curves as in Figure 3.3 that the ceiling, warm cavity and deck itself make a significant contribution to the insulation, and there is a danger that interstitial condensation will occur in the deck if insufficient insulation is provided on top of the vapour barrier; in 'warm' roof construction of this type problems generally arise because the insulation beneath the roof covering is designed to conform with Building Regulations thermal requirements and the risk of interstitial condensation has not been assessed. Examples of failures are given in section 4.7.

Calculations on the thermal properties of floors can be as complex as for roofs, and it is best to use information for typical structures, as in Table 3.6, adjusted to incorporate differences in the structure. The technique is always to work in thermal resistance R values, so that when a transmittance or U value is known, the reciprocal must be calculated in order to provide the R value before commencing the calculation. The change in R value due to a difference in material or thickness is then simply incorporated to calculate a revised R value of the element, the reciprocal being the U value. For example, one important feature of a floor is the covering; carpet in particular is an excellent insulation. Carpet has a k value of about 0.05, so that a thickness of 10 mm will have a thermal resistance R of about 0.2. A 3 m wide suspended floor above the ground will have a U value of about 1.05 according to Table 3.6, that is an R value of about 0.95. If carpet is laid on top, this R value increases to about 1.15, reducing the U value to about 0.87. Carpet and other floor coverings also have other thermal advantages, particularly reducing draughts by sealing gaps between boards and under skirtings, a point discussed more fully in section 3.3.

Carpet is not the only furnishing of importance in relation to the thermal properties of a building. For example, a normal wood frame single-glazed window will have a U value of about 4.3, or R value of 0.23. Closed curtains will introduce a cavity with an additional R value of about 0.12, increasing the total R value to 0.35 and thus reducing the U value to about 2.86, similar to the U value for an efficient double-glazed window. In the past, thick solid walls, heavy drapes and sawdust or shavings between joists were often used to improve thermal insulation, not necessarily achieving efficiencies comparable with modern dwellings but certainly similar to those in modern commercial buildings!

Thermal resistance R and transmittance U values can be calculated as explained previously, but calculations are difficult for complex structures; it is always more reliable to determine U values experimentally in the laboratory under standard conditions. Values can also be determined with reasonable accuracy on site, provided that there is a reasonable temperature difference across the building element that is being tested, and this technique is particularly useful when it is necessary to check the U value of a structure as part of an investigation. Heat flow through an element is proportional to the temperature difference and inversely proportional to the resistance, the thermal equivalent of the well known Ohm's Law for electric current. When an electric current passes through resistances in series, the potential between any two points is proportional to the resistance between them, or in thermal terms the temperature is proportional to the resistance. This is, of course, the origin of the graph in Figure 3.3 in which temperature is plotted against resistance, enabling the temperature to be determined at any interface in a building element. Conversely, the same type of graphs or calculations can be used in conjunction with temperature measurements at interfaces to calculate resistance values, including the total resistance or R value and thus the U value of the complete element.

For example, Figure 3.4 shows temperature and thermal resistance plots for simple wood and aluminium frames. T_i is the temperature of the interior air whilst t_i is the temperature of the inner surface of the element. Similarly T_o is the temperature of the outside air and t_o is the temperature of the outside surface of the element. It is apparent from Figure 3.4 that the temperature difference $(T_i - T_o)$ is proportional to the total thermal resistance R of the element, that is the reciprocal of the U value. However, it is also apparent that the temperature differences $(T_i - t_i)$ and $(t_o - T_o)$ are proportional to the internal and external resistances respectively, or

$$(T_i - T_o)/R = (T_i - t_i)/R_{Si} = (t_o - T_o)/R_{So}$$

therefore
$$R = R_{Si}(T_i - T_o)/(T_i - T_i)$$
$$= R_{So}(T_i - T_o)/(t_o - T_o)$$

or
$$U = (T_i - t_i)/R_{Si}(T_i - T_o)$$
$$= (t_o - T_o)/R_{So}(T_i - T_o)$$

Table 3.6 Thermal transmittance U values for typical floors

1. Solid floors in contact with earth

Width	U value (W/m² °C)
3 m	1.47
7.5 m	0.76
15 m	0.45
30 m	0.26
60 m	0.15

Note This table assumes that floors are square, with four exposed edges. For oblong floors the U value can be estimated from the average edge length, but f or widths of less than 10 m the narrow width becomes of increasing importance and should be considered alone at widths of 3 m or less. If one or more sides are insulated or within the building the U value is reduced by about 22% per insulated side, an adjustment must be made to appropriate sides before averaging for oblong floors, but the reduction is less for widths of less than 10 m and becomes only 14% per side at widths of 3 m.

2. Suspended wood floors above ground

Width	U value (W/m² °C)
3 m	1.05
7.5 m	0.68
15 m	0.45
30 m	0.28
60 m	0.16

Note This table assumes that floors are square and constructed with tongued and grooved boarding or similar precautions to prevent draughts from the sub-floor ventilation. For oblong floors the U values can be estimated from the average edge length, but for widths of less than 15 m the narrow width becomes of increasing importance and should be considered alone at widths of 7.5 m or less. Carpet, parquet and cork tiles can reduce U value slightly for all suspended floors, but this reduction becomes significant at narrow widths, the reduction being 2% at 15 m, 4% at 7.5 m and 6% at 3 m.

3. Intermediate floors

Construction	U value (W/m² °C)	
	Heat flow downwards	Heat flow upwards
Wood:		
20 mm wood floor on 100 mm × 50 mm joists,		
10 mm plasterboard ceiling	1.5	1.7
Allowing for 10% bridging by joists	1.4	1.6
Concrete:		
150 mm concrete with 50 mm screed	2.2	2.7
With 20 mm wood flooring	1.7	2.0

Single glazing $R = 0.181$, $U = 5.53$

Double glazing (12 mm air gap) $R = 0.354$, $U = 2.82$

Metal frame (painted or anodized) $R = 0.178$, $U = 5.62$

Metal frame (polished) $R = 0.371$, $U = 2.70$

Wood frame $R = 0.808$, $U = 1.238$

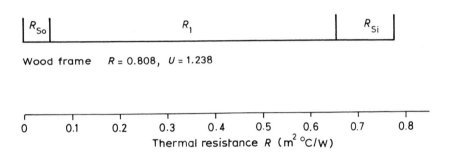

Fig. 3.4 Thermal resistance R for typical window-frame components.

These formulae enable the thermal resistance R or transmittance U values to be calculated from simple temperature measurements related to the internal or external surface resistance. The surface resistances quoted in Table 3.2 are the standard values used for design calculations. In particular, the external resistances are average values for sheltered, normal or exposed conditions which will only apply at the time of test if the wind strength happens to be the average value for such conditions. It is therefore best to avoid the use of external surface resistance and to use only the following formulae based on internal surface resistance, assessed in accordance with Table 3.7.

R and U value measurements

$$R = R_{Si}(T_i - T_o)/(T_i - t_i)$$
$$U = (T_i - t_i)/R_{Si}(T_i - T_o)$$

These formulae involve external air temperature measurements T_o; it is assumed that the external surface temperature of the element t_o depends upon the external air

Table 3.7 Internal surface resistance R_{Si} for use in R and U value measurements ($m^2\,°C/W$)

Building element	Surface emissivity	Surface resistance
Walls, windows (horizontal flow)		
bare masonry	high	0.12
white paint		0.15
anodised aluminium		0.20
polished aluminium	low	0.30
Roofs, ceilings, floors (upward flow)		
normal surfaces	high	0.11
reflective surfaces*	low	0.22
Roofs, ceilings, floors (downward flow)		
normal surfaces	high	0.15
reflective surfaces*	low	0.56

* Reflective surfaces include aluminium paint on roof surfaces and foil backing on plasterboard, even though the foil is concealed by the plaster ceiling.

temperature, which is not always the case as the external surface may be heated by the sun or cooled by evaporation of rainwater. It is therefore better to measure the actual external surface temperature t_o, but this then involves a slightly more complex calculation for R, with U determined most simply as the reciprocal of R.

R and U value measurements (improved formula)

$$R = R_{So} + R_{Si}(T_i - t_o)/(T_i - t_i)$$
$$U = 1/R$$

The external surface resistance value R_{So} in this formula is the appropriate design value from Table 3.2. This formula can be used for checking the R and U values of any building element and is particularly useful for checking thermal insulation values in individual components, such as double glazing and frame components in windows, and local features in structural elements which may form thermal bridges, such as concrete lintels over openings, and posts and beams in major buildings. In order to minimise errors, the same temperature device should be used for both surface and air temperatures. For surface measurements the probe should have a large flat collecting surface, a particularly important point when making measurements on good insulators, as there is otherwise a danger that the measured temperature will be that of the probe rather than the surface! In order to minimise this effect, the probe disc must be moved gently over the surface until a constant temperature reading is obtained, although care is necessary with some probes to avoid temperature increases through holding the probe.

3.3 Ventilation heat loss

Energy conservation tends to be directed towards minimising structural heat loss, as described in section 3.2, whether that attempt is being made by a house owner, a designer or is imposed by Building Regulation requirements. Calculations in section 3.4 will show that the structural heat loss from a typical semi-detached house, constructed in accordance with the requirements of the Building Regulations 1985, will be about 230 W/°C, although with a well insulated house this may be reduced to only about 130 W/°C; it is also necessary, however, to take account of the ventilation heat loss, that is the energy used to heat the ventilation air passing through the accommodation. The heat required for one air change per hour (1 ach) is $0.33\,W/m^3\,^\circ C$, so that in a typical semi-detached house with an accommodation volume of about $250\,m^3$, the ventilation heat loss will be about $80\,W/^\circ C$. A ventilation rate of about 1 ach is comfortable in living accommodation: reducing the rate below about 0.5 ach will generally give a feeling of stuffiness and discomfort; much higher rates are necessary when there are many people in a room or when ventilation is required for combustion, perhaps for an open fire, gas fire, or gas cooker. Although modern window and door seals can achieve much lower rates of ventilation in rooms without flues, the lowest realistic ventilation rate is about 0.5 ach, representing a ventilation heat loss in a well insulated house of about $40\,W/^\circ C$. These typical levels of ventilation heat loss are therefore very significant, representing up to 30% of structural heat loss, and even greater heat loss in houses which are draughty or fitted with open flues. Because of the need for adequate ventilation, it is difficult to reduce ventilation heat loss in conventional ways, and substantial improvements can be achieved only through suitable adaptations in building design.

Kitchens and bathrooms are often fitted with extractor fans to remove humid air and to reduce condensation in adjacent cooler areas, such as corridors and larders. When an extractor fan is operated in, for example, a bathroom it will reduce the pressure of the air in the room, causing air to flow into the room from the adjacent accommodation through the gaps around the door. This air entering the room will reduce in pressure and expand, cooling in the process; this is why the air in a bathroom feels cold when an extractor fan is working, even though the air is entering the bathroom from warm accommodation. A much better method is to lightly pressurise the air in the hall, stairwell and landing areas so that there is a flow of air through each accommodation area, including bathrooms and kitchens, which will leave the building through leaks around the windows. Slightly compressing the air will actually raise its temperature, whilst the pressure of the air will fall as it leaves the building and cause cooling. The air leaving the building is therefore cool, whereas normal ventilation air leaving a building is warm. If sufficient pressurisation is used to overcome draughts caused by wind pressure differentials on opposite sides of the building, no ventilation heat loss will occur at all, although there will be some energy consumption in pressurising the air. The air inlet is best installed, for example, in the landing ceiling so that the pressurising fan can draw air from the roof space which is warmer than the outside air. Fans are available which are suspended in a roof space to minimise noise; some of these fans are actually marketed as extractors, but they can be operated in reverse, although units designed for pressurising the accommodation air in this way are now becoming more readily available. Obviously air pressurising systems cannot operate efficiently in houses with flues, unless they are sealed off permanently or provided with efficient dampers which will seal them when the fire is not in use. Pressurising systems are also more efficient in houses with double external door systems, such as a front door leading to a porch with a second external door.

3.4 Economic factors

Table 3.8 shows the heat loss from a typical semi-detached house in watts per centigrade degree temperature difference between the outside and inside air. The difference between an older and a current house, the latter constructed in accordance with the Building Regulations 1985, is clearly apparent. Whilst it is easy to reduce the structural heat loss from an older house so that it approaches the performance of a current house, it is very difficult to achieve any significant improvement in current houses. The use of double glazing in place of single glazing may seem attractive, as it can reduce structural heat loss by about 20%, but it must

Table 3.8 Heat losses from typical houses

	Exposed area	U value	Older houses normal	Older houses insulated	Current houses normal	Current houses insulated
Roof	50 m²	1.50	75	–	–	–
		0.50	–	25	–	–
		0.35	–	–	17.5	17.5
Ground floor	50 m²	0.75	38	–	–	–
		0.60	–	30	30	30
Walls	130 m²	1.50	195	–	–	–
		0.50	–	65	–	–
		0.60	–	–	78	78
Windows	18 m²	5.00	85	–	85	–
		2.50	–	42.5	–	42.5
Structural heat loss (W°/C)			393	162.5	210.5	168
1.0 air change per hour			80	–	80	–
0.5 air change per hour			–	40	–	40
Total heat loss (W/°C)			473	202.5	290.5	208

be appreciated that this is only 20% of the reduced heat loss from a house meeting current insulation requirements so that the actual saving is not very great. To put the saving in true perspective, it is about 42.5 W/°C which is only about 11% of the typical heat loss from an older house with no special insulation, and these calculations are for very efficient double-glazed windows with air gaps of more than 15 mm and insulated frames. The replacement of existing windows is very expensive and cannot usually be justified, unless a house is very draughty and the replacements eliminate unnecessary excessive ventilation heat loss. For example, calculations several years ago indicated that the replacement of single-glazed wood frames with metal frames fitted with double glazing with a 6 mm air gap would only reduce heat loss in an average older house by about 2%, and it would take 200 years to recover the capital cost! The economics are much better for installing secondary double glazing inside an existing single-glazed wood or metal frame, or installing metal frames with thermal breaks or uPVC frames fitted with double glazing with an air gap wider than about 15 mm, as heat loss from the house can then be reduced by about 8%; if the cost of installation is taken into account, the capital cost will be recovered over about 25 years, but in the case of secondary frames installed by a houseowner the cost can be recovered over 6 years.

Whilst these calculations are based on double glazing installed in an older house, the actual reduction in heat loss and financial savings are the same, even if installation is in a modern house which is well insulated. It is apparent from these figures that advertised energy savings through insulation are often grossly exaggerated. Many advertisements and even Government publications emphasise the heat loss through a building element such as a roof, wall or window without considering whether the house already meets current Building Regulations, which limit heat loss in any case, and without reference to the total heat loss from a building or even the substantial additional loss through necessary essential ventilation, although recent revisions of Building Regulations have introduced the concept of total heat loss from a building, as well as heat gain from absorption of solar energy or leakage from plumbing and lighting. In addition, houseowners are often disappointed to discover that, whilst these theoretical savings can certainly be achieved, actual savings are much smaller because the improved insulation is resulting in higher accommodation temperatures and greater comfort rather than lower fuel consumption.

Investigations involving the thermal properties of buildings are often prompted by complaints that insulation systems are not as efficient as expected. Structural heat loss is represented by thermal transmittance or U value which can be calculated from knowledge of structural components or measured as indicated in section 3.2, but ventilation heat loss cannot be easily determined. The ventilation and floor draught effect of a normal open flue, that is, a flue which cannot be closed by a damper and which does not have an independent air supply to the hearth, is usually underestimated, as are the draughts which can be induced by a boiler, particularly if it is fitted with a forced draught system. Balanced flue systems in which a boiler fire is isolated from the accommodation can completely avoid these draught problems, but ventilation is still necessary and may need to be specially provided if a house is too free from 'natural' draughts.

3.5 Thermal movement

Thermal movement occurs when changes in temperature cause expansion or shrinkage of construction components, the main problems arising through differential movement between adjacent materials. Linear thermal expansion values for the most important building materials are shown in Table 3.9, the expansion shown being per °C. For convenience in making comparisons, all the values are shown $\times 10^{-6}$, that is, the expansion occurring per million units per °C. For example, the linear thermal expansion values for limestone and marble are 4×10^{-6}/°C, which means that these materials will expand (or shrink) by 4 units in 1 000 000 when subjected to a temperature change of 1 °C, or a movement of 80 units per 1 000 000 for a temperature change of 20 °C.

Most common building materials such as stone, brick, plaster, glass, wood, mortar and concrete have thermal expansion values within the range $4–13 \times 10^{-6}$/°C and thermal movement problems only arise with excessive lengths of material or

Table 3.9 Coefficients of linear thermal expansion (per °C)

Concrete	
gravel aggregate (1:6)	$12–13 \times 10^{-6}$
limestone aggregate (1:6)	$6–8 \times 10^{-6}$
clinker aggregate (1:6)	$8–10 \times 10^{-6}$
foamed slag aggregate (1:6)	$10–12 \times 10^{-6}$
PFA aggregate	$8–10 \times 10^{-6}$
aerated concrete	$10–12 \times 10^{-6}$
Bricks, stone	
clay	$5–8 \times 10^{-6}$
limestone	4×10^{-6}
sandstone	10×10^{-6}
granite	11×10^{-6}
marble	4×10^{-6}
slate	11×10^{-6}
Plaster	
gypsum	$10–12 \times 10^{-6}$
Wood	
hardwood and softwood	$4–5 \times 10^{-6}$
Glass	$7–9 \times 10^{-6}$
Glass fibre-reinforced polyester (GRP) sheet	20×10^{-6}
Unplasticised polyvinyl chloride (uPVC) sheet	50×10^{-6}
Metal	
aluminium	24×10^{-6}
copper	17×10^{-6}
lead	1.8×10^{-6}
steel	$10–12 \times 10^{-6}$
stainless steel	$10–16 \times 10^{-6}$

excessive temperature changes. For example, a brickwork wall subjected to a temperature rise of 20°C will expand by 0.10–0.16 mm/m (millimetres per metre), whereas a concrete feature will expand by about 0.25 mm/m. Generally, such structural materials have sufficient elasticity to accommodate such movement and no special precautions are necessary, although it must be emphasised at this stage that moisture movement, described in more detail in sections 2.3 and 4.2, is a much more serious problem and vertical movement joints are therefore essential in any case in long features. The main thermal movement problems arise where materials are subjected to excessive temperature fluctuations, particularly where joined materials have different coefficients of thermal expansion.

As accommodation temperatures are reasonably constant throughout the year, problems are usually encountered externally where materials can be subject to wide daily and seasonal fluctuations. For example, uPVC gutters fixed to wood fascia boards can often be heard creaking when periods of bright sunlight fluctuate with showers. A black gutter may have a temperature as high as 35°C, but a shower may reduce this temperature to only 5°C. The wood has a very slow response to temperature changes because of its low thermal conductivity and is virtually stable in these conditions, but the uPVC gutters have very high movement, as can be seen from Table 3.9. For the 30°C temperature change in this example, a 10 m run of gutter will move 8 mm. Similar differential movement, although only about half as great, will occur on a corrugated roof between asbestos cement sheets and glass fibre reinforced polyester (GRP) sheets, and differential movement gives rise to cracking noises in metal window frames; the movement in an aluminium frame is about three times the movement in the glass. However, the greatest differential movement arises between hot water service or central heating pipes and their supporting structures, producing cracking and sometimes groaning noises when the pipes are heating up or cooling down; the movement in copper pipes is four times as great as the movement in wood for the same temperature change, the main problem being the very large temperature change in the pipes when hot water first flows through them, typically as much as 40°C, which is very significant in relation to the relatively high movement for copper pipes.

Whilst most structural materials have sufficient elasticity to absorb the stresses and strains imposed upon them by thermal movement, there are clearly situations in which movement is significant and precautions are essential. Generally, thermal movement problems are much less serious than moisture movement problems, but there is always a possibility that extremes of temperature and moisture content will produce situations in which both forms of movement contribute. It is therefore necessary to add both the moisture and thermal movement that might occur in order to calculate the width and frequency of movement joints or other design precautions.

4

Moisture problems

4.1 Introduction

A significant moisture content is normal in all porous structural materials due to hygroscopicity, absorption by capillarity from contacting sources of moisture, or accumulation of condensation, but structural materials are only considered to be 'damp' if their moisture content is excessive, in the sense that damage can occur to materials or decorations. The level at which moisture content becomes abnormal, representing unacceptable dampness, is difficult to define, although it is usual to consider the extent of saturation of the material, as discussed in section 4.2. Fluctuating moisture contents can cause dimensional changes, as discussed in section 2.3, as well as changes in the thermal properties of materials as discussed in section 3.2. However, perhaps the most important property of moisture in building materials is the way in which the water content is an essential component in the processes of physical deterioration due to movement, salt crystallisation and freezing, chemical deterioration due to aggressive atmospheres and soils, and biological deterioration due to plants, mosses, lichens, algae, fungi, bacteria and various arthropods, particularly insects. These various deterioration processes are discussed in detail in other sections of this chapter, as well as in sections 6.4, 6.5, 7.6 and 8.6.

4.2 Moisture content fluctuations

Moisture content fluctuations cause dimensional changes which account for some of the structural movement problems already described in section 2.3. Movement with fluctuations in moisture content is greatest in wood (section 6.3), but problems can arise in other structural materials where differential movements between adjacent materials can generate very high stresses; these may lead to distortions or fractures, particularly in walls, as explained in more detail in section 11.6.

Fluctuations in moisture content can have a considerable effect upon the density of a structural material. This is particularly apparent for wood which may have a moisture content well in excess of 100% in the green state in the tree. This must be reduced by air or kiln drying to the ambient moisture content that the wood will encounter in service, normally 17% for carcassing but rather less for internal joinery, depending on the heating regime that it will encounter, so that movement is minimised and the risk of fungal deterioration is avoided. Softwood in buildings therefore has an average moisture content of 15% and a density of about 0.5 kg/l; if the wood instead has a moisture content of 85%, which is easily possible in freshly felled material, the density is increased to about 0.8 kg/l. Such large variations are not limited to wood but, in very dense materials with very low porosities, the variation in density between dry and saturated conditions is minimal. High density also means high thermal conductivity and such dense materials must be avoided where thermal insulation is required. In addition, high density is often associated with microporosity and susceptibility to crystallisation damage (see sections 4.3 and 4.4), so that very dense materials are usually avoided for general building purposes. However, lower density materials are also more porous, and if they are able to absorb excessive amounts of water their density increases, as well as their thermal conductivity; the advantages of these lower density materials are often lost if they suffer excessive water absorption.

Some years ago it was observed that the weather vane on the top of a stone church spire was swaying in the wind. Scaffolding was erected to enable the weather vane to be secured but, when the masonry of the spire was inspected, it was found that the contacting faces of the sandstone blocks were eroded and rounded, and that the entire upper part of the steeple was rocking! It was suggested by one 'expert' that the spire could be stabilised by hanging a large weight from the cap stone, but tests on stone samples showed that the compressive strength of the stone was very low, apparently through atmospheric acid pollution damage to the cementing matrix of the stone. In addition, it was found that the compressive strength when wet was rather lower than when dry, a normal variation which is often quite marked in sandstones. In this case wetting reduced the compressive strength to such an extent that the lower stones in the spire were unable to support the weight of the stone above, partly because the wetting also substantially increased the weight due to the extreme porosity of the stone. Obviously the spire was in a very precarious and dangerous condition; it was eventually repaired by the construction of a concrete core within the spire stonework.

Loss of strength with increase in moisture content is not peculiar to sandstone, but also occurs in other types of stone and in wood; if the moisture content of wood is at the fibre saturation point of 28 to 30% or above, it has only half the strength of normal 'dry' wood with a moisture content in normal ambient conditions of 12 to 15%.

The properties of air in relation to moisture content are illustrated in the psychometric diagram Figure 4.1. The 'dampness' of air is indicated by its relative humidity, that is the humidity or moisture content of the air as a percentage of the moisture content of saturated air at the same temperature. The water absorbing capacity of air increases with temperature, so that if air at constant moisture content is heated, it will appear to become drier because the relative humidity is reduced. For example, air with a moisture content of 7.5 g/kg will have a relative humidity of about 70% at 15 °C (point A), but if it is heated to 25 °C the relative humidity is reduced to about 37.5% (point B), indicating that it 'feels' drier although the moisture content is actually unchanged. Conversely, if the same sample of air is cooled to about 10 °C, its relative humidity will increase to about 100% or saturation (point C); any further cooling will result in condensation, and the temperature at which the air reaches saturation is therefore known as the dew point. An increase in moisture content will also increase the dew point, so that if the moisture content of the same air sample is increased to 13 g/kg, the dew point is increased to about 18 °C (point D). Vapour pressure is directly proportional to moisture content and does not vary significantly with temperature; the significance of vapour pressure will be discussed later in relation to interstitial condensation in section 4.7.

Air moisture content, vapour pressure and dew point cannot be measured conveniently on site, but they can be easily determined from temperature and relative humidity measurements using Figure 4.1. For example, if the air in a building has a relative humidity of about 70% at a temperature of 15 °C (point A), it can be seen that it has a moisture content of about 7.5 g/kg with a vapour pressure of about 12 mbar and a dew point of about 10 °C. Dew point is usually the most important

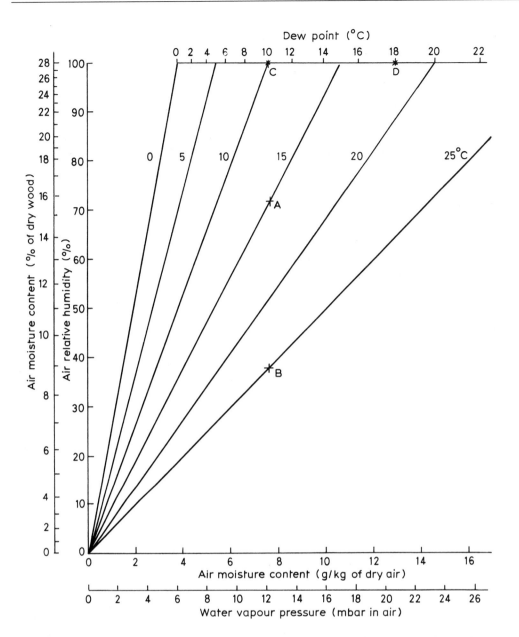

Fig. 4.1 Properties of air in relation to moisture content and temperature. This diagram enables the dew point of air to be determined by measuring the temperature and relative humidity. The average relative humidity is indicated best by the moisture content (MC) of wood in contact with the air, and wood moisture content can be used with an estimate of average air temperature to give a more accurate assessment of dew point and condensation risk. The following estimates can be used to assess risks:

Continuous heating, e.g. offices 20–24 °C, wood MC 10 ± 2%
Continuous heating, e.g. homes 12–19 °C, wood MC 12 ± 2%
Intermittent heating, wood MC 15±2%

factor, at least in winter conditions, and this information indicates that, if the air diffuses to a cooler part of the accommodation with a temperature below 10 °C, condensation will form as mist; this situation often occurs when warm humid air diffuses from a kitchen, bathroom or utility room into an adjacent cooler area such as a corridor or larder, warm air from such sources often being virtually saturated so that only slight cooling is necessary to cause condensation. If warm humid air is in contact with a cold surface with a temperature below the dew point, condensation will develop as dew, but this effect is not confined to impervious surfaces; humid air diffusing into an external wall or roof will cause condensation within the structural material as it approaches the cool exterior and reaches dew point, a phenomenon known as interstitial condensation, as explained in more detail in section 4.7.

Various instruments can be used to determine the temperature and relative humidity of air with varying degrees of accuracy. Hygrometers measure relative humidity directly, usually using the change in length of hair or a strip of paper, but generally the best technique is to use the 'wet' and 'dry' temperature method. This involves using a thermometer or other temperature measuring instrument in the normal way to determine the 'dry' air temperature and then fitting a porous material wetted with distilled or de-ionised water to the thermometer bulb or sensing element to determine the 'wet' temperature. Water evaporation from the wet material causes cooling, so that the 'wet' temperature is lower than the 'dry' temperature, except when the air is already saturated or the relative humidity is 100%; the 'dry' and 'wet' temperatures are then the same. The 'dry' and 'wet' temperatures can therefore be used, in conjunction with Figure 4.2, to determine the relative humidity. The main advantage of this method is that the same temperature measuring system is used to determine both temperatures so that any errors in the measurement virtually cancel out in normal relative humidity measurements; in conventional 'wet-and-dry bulb' thermometer instruments the two thermometers can be compared with a 'dry' air measurement and a temperature correction applied if one is reading higher or lower than the other. The method can be very accurate if there is a reasonable flow of air around the sensing element.

Temperature and relative humidity measurements are only an indication of the condition of air at the time of measurement and they must be carefully interpreted. If central heating is controlled by a clock which switches it off at night and also

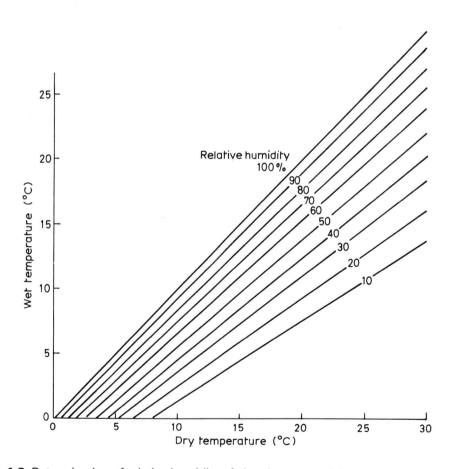

Fig. 4.2 Determination of relative humidity of air using wet and dry temperatures.

perhaps when occupants are at work during weekdays, or if other heating is used intermittently, the temperature fluctuations will affect relative humidity to a greater extent than any fluctuations in the moisture content of the air. The temperature and relative humidity measurements must therefore be used, in conjunction with Figure 4.1, to determine the moisture content of the air, and thus the dew point and the relative humidities that will occur at various other room temperatures.

If a room is well ventilated at the time of the measurements, the results may be very misleading as they will not indicate normal ambient conditions. In fact, average relative humidity can be determined much more accurately and more easily by using an electrical probe moisture meter, such as one of the Protimeter instruments, to determine the average moisture content of the softwood joinery components in the accommodation area. These measurements should preferably be made on painted wood, so that the moisture content does not change rapidly with fluctuations in atmospheric relative humidity. The most suitable sampling point is usually the top surface of the architrave across the head of a doorway in an internal partition wall, well away from possible sources of penetrating, rising or condensing moisture; the top surface is used so that the probe holes do not show! The ambient wood moisture content determined in this way can be used, in conjunction with the ambient room temperature and Figure 4.l, to determine the average air moisture content and thus the dew point and the relative humidity of the air at various temperatures. In a living room it can be assumed that the temperature during occupation usually averages about 20 °C, but there may be long unheated periods which reduce the average temperature. Generally an average of about 15 °C is appropriate for living rooms in winter and 17 °C in summer, with 12.5 °C in winter and 15 °C in summer for bedrooms, but it may be possible to estimate average temperatures more accurately by observing the apparent conditions in the building. Thus, if the wood moisture content is about 15%, this indicates an ambient relative humidity of about 70%; if the measurement is made in a living room in winter with an average temperature of about 15 °C (point A), the ambient air moisture content is evidently about 7.5 g/kg so that relative humidities at various temperatures can be determined, as well as the dew point which is about 10 °C for this example. Dew or surface condensation will develop on internal surfaces if their temperatures are below the dew point for the accommodation air; the surface temperatures will depend on the interior and exterior air temperatures in relation to the thermal properties of the roof, wall, window or door structure involved, as explained in section 3.2. Figure 4.3 enables the internal surface temperatures to be deduced from the thermal resistance R or transmittance U values for the structural component in conjunction with the internal or external temperatures, or alternatively the dew point can be used in conjunction with this diagram to determine the external temperatures below which condensation is likely to develop on internal surfaces.

Measurements of moisture content of joinery remote from any source of dampness also has a more direct value in assessing dampness in buildings as it represents, of course, the ambient wood moisture content for the room. The moisture content of the rest of the joinery in the room can then be checked to detect zones of higher moisture content; this may indicate sources of dampness and perhaps a risk of fungal development, both because dampness encourages the development of fungal decay and also because fungal activity will actually generate moisture and cause dampness in affected wood. It is explained in section 6.4 that fungal infections will only develop if the wood moisture content exceeds about 22%, the type of fungal infection depending on the moisture content and wood species. If fungal infection has developed, however, and the moisture content has since declined, the Dry rot fungus can survive by producing moisture through the digestion of the affected wood, so that areas of higher moisture content may be an indication of Dry rot activity in an otherwise dry building. This situation arises typically in buildings

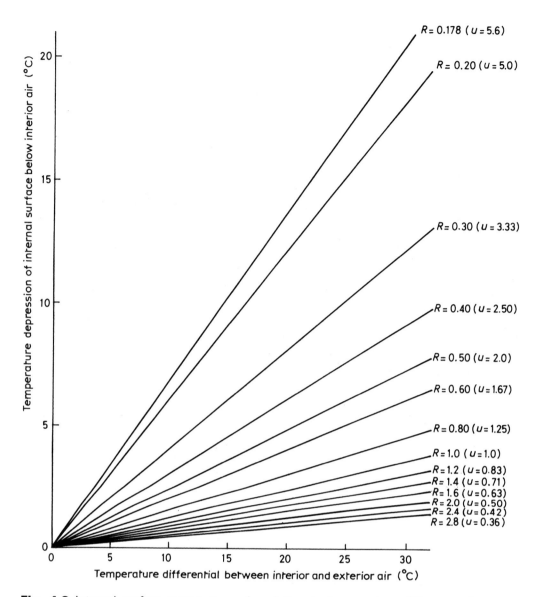

Fig. 4.3 Internal surface temperatures in relation to temperature differences between interior and exterior air. This diagram can be used to estimate internal surface temperatures for various interior and exterior air temperatures. Alternatively temperature measurements can be used to determine the R (for U) value of an external wall, door or window component. It is assumed that the internal surface has high emissivity (it is not polished or reflective) with a surface resistance of 0.123, and the external surface resistance is 0.055 (high emissivity, normal exposure). $R = 0.178$ ($U = 5.6$) therefore represents minimum thermal resistance, appropriate to single glazing and painted or anodised metal window frames.

which have been subjected to high moisture contents through leaks or indeed the processes of new construction, and where restricted ventilation has resulted in very slow drying through the critical moisture content of about 25% at which Dry rot spores germinate and develop.

Wood moisture contents are in equilibrium with atmospheric relative humidity, as explained previously. A wood moisture content of about 28% is equivalent to air saturation or a relative humidity of 100%, as shown in Figure 4.1. Higher wood moisture contents are evidence of absorption of liquid water and a clear indication of excessive dampness. Generally high moisture contents develop in this way in wood components affected by roof or plumbing leaks, or by penetrating or rising dampness, but high moisture contents can also occur through interstitial condensation, as explained later in section 4.7. Other solid materials do not behave in the same way as wood, as they are far less hygroscopic, usually achieving moisture contents of less than 5% when in equilibrium with saturated air, that is air with a relative humidity of 100%. High moisture contents in plaster, mortar, brick or stone are therefore clear indications of excessive dampness, but the accurate measurement of moisture content of such materials is surprisingly difficult. The most accurate technique is to take a sample and to seal it in a moisture-proof container, and later to weigh it, dry it in an oven, weigh it again, and then calculate the dampness from the weight loss in comparison with the dry weight. This oven-drying method is obviously too cumbersome for use in connection with routine building inspections and can be considered realistic only where a special investigation is being made. A more convenient method on site is to mix the sample with calcium carbide which will react with the water content to liberate acetylene gas; if this reaction occurs in a 'bomb' or sealed pressure vessel, the pressure will give a measurement of the amount of water present in the original sample. This is the principle of all carbide moisture detectors. Samples are normally obtained by drilling, usually with a tungsten-tipped bit with a diameter of about 9 mm ($\frac{3}{8}$"); high-speed drilling and blunt drills must never be used as they may result in overheating and the loss of moisture from the sample. Usually about the first 12 mm ($\frac{1}{2}$") of each drilling is kept separate and its moisture content compared with deeper samples; the deeper samples give the true moisture content of the wall and higher moisture contents in surface samples suggest condensation or hygroscopic salts. If hygroscopic salts are suspected, a surface sample can be conditioned in a shallow dish or watch glass over a tray of water in a container at 75% relative humidity, achieved most easily using a saturated solution of common salt (NaCl), the hygroscopic moisture content after overnight exposure indicating the dampness that can be attributed to the presence of hygroscopic salts. Thus if the moisture content of the sample is, say, 8% and the hygroscopic moisture content is also 8%, the dampness in the sample can be attributed entirely to the presence of hygroscopic salts, but if the normal moisture content is, say, 12% relative to a hygroscopic moisture content of 8%, the extra dampness can be attributed to an external source such as penetrating, rising or condensing dampness; if the surface samples have a higher moisture content than the deeper samples but the hygroscopic moisture content is low, the higher moisture content can be attributed to condensation.

Carbide instruments can give accurate measurements of moisture content but a warning is necessary; some instruments such as the 'Speedy' give moisture content as a percentage of wet weight, thus giving a rather low result in comparison with the more normal percentage of dry weight. However, the readings can be easily converted:

$$\text{moisture content} = x/(100 - x)\% \text{ of dry weight}$$

where x is moisture content as % on wet weight.

Thus, if the moisture content is 20% of wet weight, this is equivalent to 25% of dry weight.

The actual moisture content of a building material is only significant, and the material is only considered to be damp, if the moisture content is excessive, that is if it causes damage to materials or decoration; actual moisture content is therefore of little significance and it is more important to consider whether a material appears to be damp or dry. If a measurable moisture content is found, say 5%, the significance of this figure is difficult to assess. If the total porosity of the material is 5%, it is clearly saturated, but if the total porosity is 40%, it is relatively dry. It is also necessary to distinguish between total porosity and absorption, the latter indicating the amount of water that can be absorbed under normal circumstances, whereas total porosity indicates the amount of water that the material will absorb when vacuum impregnated; a material that is naturally saturated by absorption will often contain considerable amounts of trapped air, the absorption relative to saturation being known as the saturation coefficient. For these various reasons it is best to avoid all comments on the significance of particular moisture contents unless the measurements are made on wood, and for measurements on other materials to be used only to establish moisture content gradients to diagnose sources of dampness. The use of electrical conductivity moisture meters is often criticised as the relationship between conductivity and moisture content is rather inconsistent, but it will be appreciated from these comments that even accurate measurements of moisture content will give little information on significant 'dampness' in building materials, and for diagnostic purposes on site an electrical conductivity moisture meter is considerably more convenient than oven drying or carbide methods. The main criticism of electrical conductivity meters is the way in which the readings are exaggerated by the presence of soluble salts, but many of the salts encountered in building materials are hygroscopic, as explained more fully in section 4.10, and tend to absorb moisture from the air in any case.

It will be apparent from these comments that it is recommended, when inspecting a building for dampness, to establish first the ambient conditions within an accommodation area by measuring the moisture content of painted joinery in a position which is unlikely to be affected by any external source of dampness, such as the top architrave of a doorway in an internal partition. This moisture content will permit the ambient conditions within the accommodation to be established, and also enable measurements on other joinery components to be put in perspective; areas of excessive moisture content, generally wood moisture contents above 18%, can then be readily detected. Measurements on other structural materials are best avoided, except as comparative measurements designed to detect moisture gradients and thus locate sources of dampness.

Although fungal decay in wood components is certainly the most serious consequence of excessive moisture contents in building components, other materials such as paint, wallpaper and particularly gypsum plaster will also deteriorate. In addition, dimensional changes with moisture content fluctuations can cause particularly serious problems, as discussed in section 2.3; for many materials, including wood and cement products, shrinkage on initial drying is much greater than the fluctuations that occur in normal service.

4.3 Frost damage

When water is cooled below its freezing point 0 °C it begins to solidify and form ice. The density of ice is less than that of water so that ice floats on water, conversely the volume of ice is greater than the water from which it is derived. Freezing therefore involves expansion which can cause rupturing of porous materials. The most common form of damage is progressive erosion of a surface due to fracturing of the pore structure. If the surface is densified through some other effect, such as the formation of calcium sulphate in the surface of a limestone subject to a polluted atmosphere, the stone immediately beneath this densified surface is often more porous and weaker, and severe frost can cause the entire densified surface to spall away to a depth of 10 mm or more.

Obviously some materials are much more susceptible to frost damage than others, and these differences can be attributed to the amount of water present when the material freezes relative to the total porosity or available space. When a porous surface is in contact with liquid water, such as rainfall or dampness rising from the soil, the amount of water that can be absorbed is not necessarily sufficient to saturate the porosity, leaving air spaces which can accommodate some of the expansion as ice forms and therefore minimise rupture damage. The amount of water absorbed by a porous material relative to its porosity is known as the saturation coefficient, which is 1 if the amount of water that is normally absorbed is equal to the total porosity of the material. Low saturation coefficients therefore indicate porous materials that are likely to have good resistance to frost damage. However, whilst the saturation coefficient can give a good guide to comparative durability within stones from particular beds or within closely similar types, the method cannot be used to compare different porous materials, such as bricks with sandstones, or even a limestone with a different type of limestone.

A porous solid will generally absorb water by capillarity, the absorptive forces being greater for smaller pores. In addition, water disperses more readily from larger pores in which ventilation is obviously more efficient. For these reasons small pored or microporous materials are likely to absorb water more strongly and to disperse it subsequently less efficiently; they are therefore more likely to have a high moisture content during freezing conditions and thus more likely to suffer frost damage. One method for assessing the probable durability of a porous material is therefore to assess pore size distribution. Usually this is achieved most simply by defining a critical pore diameter, such as 0.005 mm, micropores being smaller and macropores being larger than this size. A sample of the dry porous material is saturated with water by vacuum impregnation to measure the total porosity, and then a critical pressure or vacuum is applied which will remove the water from the macropores alone, capillary forces retaining the water in the smaller micropores.

In the classical method, samples of saturated material are placed in contact with a suitable capillary bed and subjected to a vacuum of 600 cm of water, equivalent to a pore diameter of 0.005 mm and sufficient therefore to evacuate macropores larger than this size. The preparation of the samples is difficult and the test consequently expensive. However, a method was developed many years ago in the author's laboratory in which the small sample of material is replaced by a hole drilled in a random shaped block. A probe is sealed into the hole using an expanding cuff and water injected until a constant flow rate indicates saturation, this flow rate being a measurement of the total porosity of the material. Air is then introduced at a pressure which will remove water from the macropores alone, the air flow rate being a measurement of the macroporosity. In this way it is possible, after taking the relative viscosities of air and water into account, to determine macroporosity as a proportion of total porosity, thus indicating the probable durability of the material. In fact, all these methods only indicate the probable durabilities of stones of average porosity, as resistance to freezing expansion also depends on the cohesiveness of the stone. Thus a stone of low porosity is likely to be much more durable than a stone of very high porosity, and these variations must be taken into account when assessing pore-size distribution results.

While saturation coefficient and pore-size distribution tests can be used in a laboratory for routine durability assessments, they must be matched with other tests which will closely reproduce failure in service. In many countries freeze-thaw cycles are used to assess resistance to frost damage, but the results are not very consistent. Precisely the same type of rupture damage occurs through the growth of crystals of various salts, as explained in section 4.4, and more consistent results can be obtained using crystallisation of magnesium or sodium sulphate during a wetting and drying sequence, the results normally being assessed by the weight loss that occurs through surface deterioration; in the United Kingdom the most widely used test involves 15 crystallisation cycles with 14% sodium sulphate decahydrate.

Since about 1988 porosity tests have been used in the author's laboratory to calculate a durability factor in which cohesive strength is represented by the reciprocal of porosity. Several alternative methods of calculation designated A to H were

Table 4.1

Durability class	Durability factor D	Crystallisation (% loss after 15 cycles)
A	<4	<1
B	4–5	1–5
C	5–7.5	5–15
D	7.5–12.5	15–35
E	>12.5	>35

Note Class F, which indicates fracture of a test block before completion of the crystallisation test, is indicated by a factor D in excess of about 25.

postulated, but factor D was found to match most closely assessments using salt crystallisation. This method is now used routinely for the rapid durability assessment of porous stone, with sodium sulphate crystallisation used as a confirmatory method when assessing stone from a new source. Durability factor D is calculated from the square of capillarity divided by porosity, or if using old tables it is the square of saturation coefficient times porosity (assuming that porosity is expressed as % volume; if porosity is % mass, divide it by density). For limestones and sandstones of normal porosities durability factor D can be closely matched with 14% sodium sulphate crystallisation loss after 15 cycles, as given in Table 4.1.

Resistance to frost damage depends not only on the properties of the porous materials involved but also on the type of exposure. For example, stone used for smooth surface ashlar walls, properly protected by overhanging eaves and other details, will be much more resistant to frost damage than stone in very exposed situations such as copings, string courses and pinnacles, or stone that is continuously absorbing water from the soil at the bases of walls or as pavings or steps. Thus the most durable stone must be used in these particularly exposed situations, whilst less durable stone can be used for protected ashlar or interior work.

Caen limestone traditionally has a good reputation for durability and it has lasted well in major buildings, such as Winchester Cathedral. However, the demand for stone during the late nineteenth and early twentieth centuries resulted in the working of beds which were much more accessible and therefore cheaper, but also less durable. Victoria College in Jersey is an example of a major building which was constructed using this less durable stone, but erected in a climate in which severe frosts are not very frequent. As a result the stone survived for several years free from defect until a severe frost caused spalling damage, and this infrequent cycle of damage has continued ever since. Some of the periods of damage have been followed by remedial works comprising plastic repairs, that is the use of a matching mortar to replace spall damage, and these repairs also appeared to be sound until the next severe frost which actually spalled the stone beneath the repairs! In more severe climates with frequent frosts it is impossible to follow the sequence of deterioration in this way and to positively identify frost as the cause. In urban areas the damage caused by atmospheric pollutants may be more severe than that caused by frost, as explained more fully in section 4.4, and the situation is therefore even more confused, particularly as salt crystal growth causes the same damage to stone as freezing expansion.

4.4 Chemical damage

Chemical damage depends, as does frost damage, on the presence of moisture, and it is therefore appropriate to consider it in detail in this chapter. This section is concerned with chemical damage to non-metallic materials, as metal corrosion is considered separately in section 4.5.

The simplest form of chemical damage is rainwater leaching of soluble materials. Most structural materials have excellent resistance to leaching, although occasionally rainwater leaching problems arise, usually on statuary rather than on structural components, where alabaster (gypsum or calcium sulphate) or stones containing magnesium carbonate have been used; both these compounds are appreciably soluble in rainwater. However, air always contains carbon dioxide, generated by respiration and combustion, which dissolves in rainwater to produce carbonic acid which reacts with carbonates to form more soluble bicarbonates. Whilst this action can cause slow erosion of porous limestones and sandstones dependant on carbonate cementing material, and even etching of marble surfaces, the soluble bicarbonates are unstable and revert on drying to carbonate so that the effect is mainly a redistribution of carbonate rather than severe loss.

The intensity of the damage is much greater in polluted urban atmospheres. The most common pollutants are oxides of nitrogen, particularly associated with coal combustion, and oxides of sulphur, associated with both coal and oil combustion. The normal oxides in the atmosphere are nitrous oxide and sulphurous oxide, which dissolve in rainwater to form nitrous and sulphurous acids which react with carbonates to form nitrites and sulphites respectively. However, nitrites and sulphites are never found on analysis of eroded limestones in urban atmospheres, but instead nitrates and sulphates are found, indicating that the acids have been oxidised at some stage to the much more aggressive nitric and sulphuric acids. The explanation for these observations is the presence on the stone of nitrating and sulphating bacteria, as discussed in more detail in section 7.6.

Nitric and sulphuric acids can cause rapid erosion of carbonate and even slow erosion of silica components in stone, mortar and concrete. The rate of erosion obeys the normal chemical law of mass action, that is the rate of reaction depends on the concentration of the reacting components. In the case of porous building materials, this law can be interpreted as the surface area of the solid components in contact with the aggressive solution, so that macrocrystalline stones, composed of large crystallites, erode less rapidly than microcrystalline stones which present a much greater active surface area in contact with the aggressive solution. Similarly, lower porosity leads to slower leaching than higher porosity, but pore size distribution is also important. Microporous stone, which absorbs rainwater and dissolved pollutants strongly by capillarity and which limits subsequent ventilation drying, will erode much more rapidly than a macroporous stone; this will suffer much shorter periods of progressive deterioration, mainly because it remains wet for shorter periods. For example, a Bath limestone, such as Monks Park, is much less durable than a Portland limestone, such as Whitbed, despite their similar total porosity, because the Bath stone is microcrystalline and predominantly microporous, whilst the Portland stone is macrocrystalline and predominantly macroporous.

Acid pollution erosion of calcium carbonate limestones results in the formation of calcium nitrate and calcium sulphate, or magnesium nitrate and magnesium sulphate if magnesium carbonate is present. Limestones containing pure magnesium carbonate or even simple mixtures of calcium and magnesium carbonates are not common, but magnesian limestones containing double carbonates of calcium and magnesium can give very good resistance to erosion if they are in the form known as dolomite; this is the explanation for the good erosion resistance of the magnesian Huddlestone limestone used for the more exposed features at York Minster. However, direct acid erosion is, unfortunately, only half of the deterioration story, as the salts that result from this erosion accumulate at the surface through water evaporation and crystallise, the crystals expanding and causing rupture damage similar to that caused by frost, as described in section 4.3. The degree of expansion of the crystals depends on the amount of water that is absorbed by the salt on crystallisation, and this depends in turn on the nature of the carbonate present in the stone and the pollutant acid to which it has been exposed:

$CaSO_4.2H_2O$
$Ca(NO_3)_2.4H_2O$ (hygroscopic)
$MgSO_4.7H_2O$
$Mg(NO_3)_2.6H_2O$ (hygroscopic)

All four salts can cause crystallisation erosion, but this is most severe in the case of the magnesium salts because of their high water of crystallisation and thus their considerable expansion on crystallisation. The nitrates are hygroscopic and will absorb moisture from the air, their presence in plaster on a chimney breast being the usual explanation for persistent damp patches. Calcium sulphate is less soluble than the other salts and can accumulate in the surface of porous stone to cause surface densification; the area beneath the densified surface is weakened by the acid attack and, if the stone freezes in a saturated condition, the densified surface may spall away to a depth of 10 mm or more. If acid attack occurs on crystalline marble, the salt crystals develop between the marble crystals, causing detachment of the surface crystals and the exposure of a rough crystalline surface beneath, an effect usually known as 'sugaring'. In addition, the marble will expand due to salt crystallisation between the carbonate crystals, and in polluted areas subject to a prevailing wet wind this expansion will be concentrated on one face of the marble and will result in bending or torsion, an effect often seen in Carrara marble headstones in urban cemeteries.

Salt crystallisation damage is not caused only by salts generated within a stone by pollutant acids, as salt solutions may be introduced from other sources. The most serious problems arise when a normally durable stone is exposed to washings from another stone, subjecting it to unusual deposits of crystallising salts. Pure siliceous sandstones, that is stones comprising silica particles with a silica cementing matrix, are generally exceptionally durable, although their cohesive strength is not usually very great; the durability arises mainly from the excellent resistance of silica to pollutant acids and thus freedom from any generated salts. However, when stones of this type absorb washings from limestones, calcium or perhaps magnesium salts are introduced which accumulate at the evaporating surfaces; these can cause severe erosion problems in sandstones with durability reputations or assessments which do not take account of such salt crystallisation. This effect is common where washings from limestone walls accumulate on sandstone pavings. In urban areas where sandstones are extensively used, such as Edinburgh and Glasgow, erosion of the sandstone is often seen beneath each horizontal mortar joint through salts generated in the mortar, a synthetic carbonaceous sandstone, washing into the adjacent sandstone. Washings from magnesium limestones can cause similar problems in calcium limestones because of the unanticipated expansion of the magnesium salts. This was a particular problem for many years at York Minster where the progressive erosion damage was attributed to the inherent poor durability of the original magnesian limestones; repairs were therefore carried out using calcareous limestones with excellent durability reputations, but these calcareous limestones eroded rapidly when exposed to magnesian limestone washings, indeed much more rapidly than the magnesian limestone which it had replaced and which had only eroded to a limited extent over many centuries of exposure.

Erosion damage can also occur in bricks through the presence of sulphates derived from the clay; these sulphates can cause efflorescence as well as sulphate attack of the mortar, as described in section 7.4. Salts can also be introduced from other sources, such as in rising dampness; erosion just above a ground line is generally associated with freezing in continuously saturated condition or crystallisation of absorbed salts such as sodium sulphate which causes particularly severe erosion because it has ten molecules of water of crystallisation:

$$Na_2SO_4.10H_2O$$

It is, of course, this exceptional crystalline expansion that is the reason for the choice of this salt in standard testing of resistance to crystallisation stress from salts or frost, as described in section 4.3.

Damage due to pollutant acids is not restricted to the porous limestones and sandstones, or even to stones containing carbonates. Sugaring expansion damage to dense crystalline marbles has already been described, but similar damage can cause delamination in poor quality slates, and even polished granite surfaces will be etched to a certain extent by pollutant acid attack.

It is often imagined that pollution damage has reduced over the years due to the reduction in coal combustion and the introduction of clean air legislation. Whilst it is true that particulate pollution and oxides of nitrogen have become progressively less significant, mainly as a result of reduced use of coal but also through improved combustion and electrostatic precipitators in, for example, coal-fired power stations, pollution due to oxides of sulphur has become progressively more severe. This increase in sulphur pollution can be attributed partly to the increasing use of heavy fuel oil for heating industrial and commercial premises, as this oil is relatively crude with a high sulphur content, but it is also partly due to the legislation which limits only the concentration of sulphur in effluent gases, rather than the total amount of sulphur emitted; if the sulphur content is too high it is only necessary for the operator to blow air into the flue to reduce the emitted concentration, thus distributing the pollution over a wider area. This is obviously a problem that will only be remedied by introducing legislation to restrict the amount of sulphur in fuels.

Wood also suffers chemical damage in some circumstances, but this is discussed in detail in section 6.6.

4.5 Metal corrosion

The corrosion of iron and steel is a form of oxidisation which commonly occurs in the presence of air and moisture when it is known as 'rusting'. Oxides formed on other metals, such as aluminium, tend to form a protective layer rather than being part of a corrosion process. However, these processes and their significance can only be properly understood through a basic knowledge of electrochemistry.

Salts typically ionise when dissolved in water, or separate into positive and negative charged ions known respectively as cations and anions:

$$NaCl = Na^+ + Cl^-$$
$$NaOH = Na^+ + OH^-$$
$$CaCl_2 = Ca^{++} + 2Cl^-$$

Water itself also ionises, provided that it is not absolutely pure and other ions are present:

$$H_2O = H^+ + OH^-$$

If an electric current is passed between electrodes in an electrolyte or solution, the negatively charged anions will move towards the positively charged anode, and the positively charged cations will move towards the negatively charged cathode. If the electrolyte consists of water with only sufficient acid to cause ionisation of the water, the reactions at the electrodes or electrolysis may be represented by:

Cathode reactions

$$H^+ + e = H$$
$$2H = H_2 \quad \text{(hydrogen gas)}$$

Anode reactions

$$OH^- - e = OH$$
$$4OH = 2H_2O + O_2 \quad \text{(oxygen gas)}$$

The e represents an electron or negative unit of electricity flowing from the cathode to the anode, this electrical flow causing hydrogen to be formed at the cathode and oxygen at the anode. If salts are present in significant concentrations in the electrolyte, these reactions are modified. For example, if common salt or sodium chloride is present:

Cathode reactions

$$H^+ + e = H$$
$$2H = H_2 \quad \text{(hydrogen gas)}$$

Anode reactions

$$Cl^- - e = Cl$$
$$2Cl = Cl_2 \quad \text{(chlorine gas)}$$

It might be expected that sodium and hydrogen would be generated at the cathode, but hydrogen has a lower discharge potential and therefore occurs preferentially, whilst at the anode chlorine is formed as chloride ions have a lower discharge potential than hydroxyl ions. However, with copper sulphate solution, copper ions have a lower discharge potential than hydrogen ions and copper is therefore formed at the cathode, but hydroxyl ions have a lower discharge potential than sulphate ions so that oxygen is formed at the anode:

Cathode reaction

$$Cu^{++} + 2e = Cu$$

Anode reactions

$$OH^- - e = OH$$
$$4OH = 2H_2O + O_2 \quad \text{(oxygen gas)}$$

The formation of copper at the cathode is, of course, the basis of electroplating.

If a metal electrode is placed in an electrolyte, an electric potential develops between the electrode and the electrolyte. These electrode potentials are usually referred to the potential for hydrogen which is considered to be zero, and electrode potentials therefore vary from 'base' metal, such as lithium at -3.02 volts, to 'noble' metal, such as gold, at $+1.42$ volts, as shown in Table 4.2. If electrodes from different positions in this electrochemical series are placed in an electrolyte, the potential between the electrodes will be equal to the difference between their electrode potentials as shown in the table, and this is the basis of an electric cell. The passage of current between electrodes usually results in the anode dissolving and ions from the electrolyte being deposited at the cathode, and it is this dissolving material at the anode which is the essential feature of most corrosion processes involving the establishment of electric cells and current flow.

Table 4.2 The electrochemical series

Metal	Ion	Electrode potential (volts:
'Base' end		
Lithium*	Li^+	−3.02
Potassium*	K^+	−2.92
Sodium*	Na^+	−2.71
Magnesium*	Mg^{++}	−2.38
Aluminium*	Al^{+++}	−1.67
Zinc	Zn^{++}	−0.76
Chromium	Cr^{++}	−0.71
Iron	Fe^{++}	−0.44
Cadmium	Cd^{++}	−0.40
Nickel	Ni^{++}	−0.25
Tin	Sn^{++}	−0.14
Lead	Pb^{++}	−0.13
Hydrogen	H^+	0.00
Copper	Cu^{++}	+0.34
Mercury	Hg_2^{++}	+0.80
Silver	Ag^+	+0.80
Platinum	Pt^{++}	+1.20
Gold	Au^{+++}	+1.42
'Noble' end		

Notes The values for the metals marked * are theoretical; aluminium gives nobler values due to the formation of an oxide film, and the other metals evolve hydrogen which interferes with measurements.

Some metals and alloys will corrode naturally, the corrosion product accumulating on the surface and becoming thicker with time, often inhibiting further corrosion. The formation of oxide on the surface of aluminium inhibits corrosion in this way, and corrosion only occurs when the protective oxide film is damaged, perhaps by chemical attack. Corrosion damage is usually due to electrolytic action or the formation of cells and is therefore dependent on the presence of moisture. Although a cell formed in this way will result in the anode metal being dissolved, it will also result in the formation of hydrogen at the cathode; this formation of a hydrogen film will progressively reduce the current flow until eventually the cell will become polarised and stable without current flow or corrosion from the anode. However, if there is a 'depolariser' present which can oxidise the hydrogen, the film will be removed or will not form, and rapid corrosion of the anode will occur; oxygen from the air dissolving in the electrolyte is the most common depolariser and thus the cause of rusting on unprotected steel, but there are other less obvious causes of depolarisation. For example, if an iron or steel object such as a pipe is buried in mud or clay, the conditions are anaerobic or free of oxygen, and polarisation with formation of a protective hydrogen film readily occurs. However, if the mud contains sulphates, sulphate-reducing bacteria may be present, such as *Desulfovibria* species which remove the protective film of hydrogen, greatly accelerating the corrosion process; these bacteria are also involved in the sulphate cycle described in section 8.6. This anaerobic corrosion process can cause very severe damage to unprotected gas, oil, water and sewage pipes, and can also cause catastrophic collapse in steel structures subject to anaerobic conditions, such as stagnant water in hollow piles or the hollow legs of drilling rigs and oil production platforms, and even in situations where careless welding results in hollow spaces within the welds. It is therefore worth considering the processes that occur in this way.

The initial corrosion reactions on iron or steel are:

$$2H_2O = 2H^+ + 2OH^-$$

Anode reaction

$$Fe = Fe^{++} + 2e$$

Cathode reactions

$$2H^+ + 2e = 2H = H_2 \qquad \text{(hydrogen gas)}$$

The film of hydrogen formed on the cathode in this way prevents further current flow and inhibits dissolution of metal or corrosion; however, if sulphate-reducing bacteria are present and active, they utilise sulphate ions as electron acceptors in respiration, depolarising the cathode and removing the protective film of hydrogen in the process, causing dissolution of iron at the anode and the formation of ferrous sulphide and hydroxide:

$$SO_4^{--} + 8H = S^{--} + 4H_2O$$
$$Fe^{++} + S^{--} = FeS$$
$$Fe^{++} + 2OH^- = Fe(OH)_2$$

These reactions may be summarised:

$$4Fe + SO_4^{--} + 4H_2O = FeS + 3Fe(OH)_2 + 2OH^-$$

The ratio of ferrous sulphide to hydroxide is a useful diagnostic feature which can be used to confirm the involvement of sulphate-reducing bacteria in corrosion. The process is illustrated diagramatically in Figure 4.4.

Metal corrosion can be prevented in various ways. When metal is permanently in contact with moisture forming an electrolyte, one method for inhibiting corrosion is to prevent the metal from becoming an anode and thus prevent dissolution of metal.

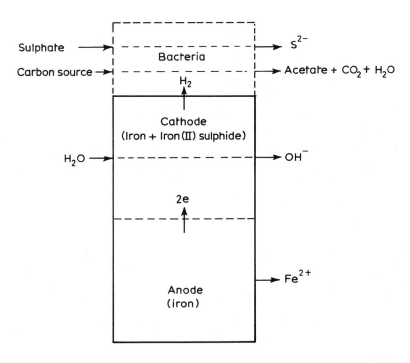

Fig. 4.4 Diagrammatic representation of the anaerobic corrosion of iron.

In an impressed current system, the metal requiring protection is connected to a source of power and a second electrode, the applied potential then ensuring that the metal remains the cathode in the circuit, the second electrode becoming the anode but being free from corrosion provided that it is further towards the noble end of the electrochemical series; platinum is often used for such electrodes, usually in the form of platinised titanium in order to minimise cost. Another method is to attach a sacrificial anode to the metal requiring protection. Generally such anodes are used on iron or steel and are manufactured from alloys of magnesium and aluminium, being anodic as they are towards the base end of the electrochemical series relative to iron. In this system the attached anode corrodes steadily instead of the metal requiring protection, but sacrificial anodes require replacement at regular intervals. Both these systems are known as cathodic protection as they cause the protected metal to become the cathode in the system. The most serious corrosion problems occur when two connected metals form a cell in an electrolyte, such as copper and iron fittings on a boat immersed in sea water, but cathodic protection is still effective provided that the anode of an impressed current or sacrificial system is lower down the electrochemical series than the metals requiring protection, that is further towards the base end of iron in the case of a mixture of iron and copper.

These descriptions of corrosion have referred, for simplicity, to metal immersed in sea water and other corrosion hazard situations, but similar corrosion can also occur in buildings, as well as in exposed metal structures such as fences, as described in section 17.4. For example, service water tanks and pipes may corrode under aerobic conditions, but central heating and sewage systems may corrode under anaerobic conditions, as described in section 16.3. Metal fixings such as masonry dowels, cramps and cavity wall ties will corrode unless suitable precautions are taken, such as the use of adequately galvanised steel or stainless steel alloys. Steel reinforcement in concrete will corrode in various ways, as described in section 8.5, including the formation of cells; these cells do not involve different electrodes but instead involve different electrolytes or different concentrations of the same electrolyte in contact with a single piece of metal which becomes a cathode in one area and an anode elsewhere, with corrosion at the anode. Whilst corrosion ultimately involves loss of strength through loss of metal, the formation of corrosion products such as rust results in expansion which may rupture the structure and represent a more immediate and more severe problem.

Paint coatings are often used to reduce corrosion in metals exposed to the weather, but simple coatings are only effective if they remain completely intact. The main protection against corrosion of iron or steel is always polarisation or the formation of a hydrogen film but, if a coating is damaged and metal is exposed to the atmosphere, oxygenation and depolarisation will occur, and a potential difference will be established between the parts of the metal surface which are oxygenated and those which do not have access to oxygen. This differential oxygenation explains the rusting of iron preferentially under areas which are covered by a coating or even wet rust; whilst this is the most obvious explanation for preferential corrosion beneath a coating surrounding a damaged area, sulphate-reducing bacteria also contribute to corrosion in anaerobic conditions beneath coatings. For these reasons oxidising pigments such as red lead, lead chromate and zinc chromate are generally most effective as coating primers, zinc chromate being used most widely, partly to avoid the toxicity generally associated with lead compounds but also because it is more effective, presumably because the zinc also acts as a bactericide.

Various types of metal coatings or platings are also used to protect steel. Tin plating only remains effective provided the tin covering remains completely intact, as tin is further up the electrochemical series and therefore more noble than iron. When damage occurs to a tin coating and the damaged area is covered by moisture forming an electrolyte, the exposed iron is anodic and current flows through the electrolyte to the tin, resulting in dissolution and corrosion of the exposed iron. However, if the tin plating is replaced with zinc, which is more base than iron in the electrochemical series, the current flows in the opposite direction so that, when damage occurs to galvanised steel, the zinc is dissolved but plates out on the exposed iron, repairing the damage and providing new protection. Obviously this protection of any exposed iron through sacrificial corrosion of the zinc is an advantage, although it must be appreciated that zinc corrodes slowly when exposed to the weather, and much more rapidly in polluted urban atmospheres or when subject to sea spray, so that the protection offered by galvanising and sheradising depends on the thickness of the deposited zinc, as well as the corrosive conditions to which it is exposed.

Lead is used traditionally where corrosion resistance is required. When exposed to air lead tarnishes rapidly through formation of a superficial film of hydroxide and carbonate, and this film inhibits further deterioration and provides the corrosion resistance for which lead is well known. However, natural soft water or even water from a water softener is slightly acidic and can dissolve this protective film and allow lead corrosion; this is why lead pipes are no longer permitted for water supplies.

When used as a roof covering or as flashings or damp-proof courses lead is normally very corrosion resistant in non-polluted atmospheres. However, leaching from fresh mortar will cause light coloured lime deposits on lead but, whilst these are unsightly, they are only superficial and do not cause corrosion. However, flue gases combined with rainfall can cause corrosion; this may blend with lime deposits to give some yellow staining, typically on flashings close to chimneys, although corrosion damage is not usually significant if lime deposits neutralise the acid.

The most serious problem with lead involves water discharge from roofs and walls affected by moss and particularly lichen growth, as they produce organic acids which can cause severe lead corrosion. The damage is most severe when the growth is on slate, some sandstone and granite, but much less severe if the growth is on limestone which partly neutralises the acid. It is always sensible, particularly with roofs affected by moss and lichen growth, to protect lead roofs and gutters receiving discharge with a bitumen or tar coating to reduce this corrosion.

Stainless steel is often used where corrosion resistant metal is required; it is even used today as an inexpensive alternative to lead or copper as a roof covering. However, various stainless steel alloys are available. Type 304 is widely available and inexpensive, and suitable for use in most external conditions, but it has poor resistance to chloride and will corrode if subject to sea spray or used, for example, for handrails and other fittings for swimming pools using chloride hygiene systems; type 315 is better and type 316 is best in avoiding chloride damage, rust staining and pitting corrosion.

4.6 Leaks and floods

Building designs must take account of normal sources of moisture such as penetrating, rising and condensing dampness, but excessive moisture may arise in other ways. Leaks and floods represent accidental events which will not have been anticipated in the design, and which may therefore require particular attention when they occur.

Dampness may arise through penetration into an apparently sound roof through extreme weather conditions, such as wind-blown rain or snow. These problems are usually encountered in older buildings lacking the sarking felt under the tiling battens which is designed to prevent such penetration, although in modern buildings penetration still sometimes occurs due to errors in design or construction. The most common problems in modern buildings occur at the eaves, through sagging of the sarking felt or failure to project it sufficiently to discharge into the gutters, so that the wallhead becomes wet, typically causing fungal decay in the wall plate and rafter feet, and perhaps also affecting window lintels and frames beneath. Dampness in this area is also caused sometimes through careless tiling or slating at the eaves; this allows water penetration, the sarking felt then providing the only barrier against water penetration, and leaks develop as the sarking felt deteriorates with time. Obviously there are many other roof construction defects that may cause similar dampness problems, as described in Chapter 13, periodic leaks being associated with certain wood destroying fungi, particularly *Poria* species. Fungal decay problems are discussed more fully in section 6.4, in which it is explained that the most serious fungal decay danger arises when wet wood begins to dry in poorly ventilated conditions, as with the lower surfaces of plates in contact with wallheads and with built in lintels and beam ends. In these conditions the moisture content will decrease very slowly, eventually passing through the critical moisture content of about 25% at which the spores of the Dry rot fungus *Serpula lacrymans* germinate. Spore germination is encouraged by acidity, but wood is acid if it has been previously attacked by some of the Wet rots that are active at higher moisture contents, such as the *Poria* species that have been previously mentioned. The conditions following leak repairs are therefore particularly conducive to the development of Dry rot; the seriousness of the situation is associated with the fact that the Dry rot fungus is able to spread widely behind plaster and through masonry in the search for more wood through the development of rhizomorphs or conducting strands. These can convey water and nutrient, but when further wood is found in poorly ventilated conditions the digestion of the wood will generate sufficient moisture for the fungus to maintain the moisture content of the wood and the humidity of the surrounding area at an appropriate level for optimum development.

The danger of Dry rot development is not, of course, restricted to roof leaks, but is associated with drying following any form of flooding, including water from plumbing leaks and fire fighting. The easiest way to avoid the Dry rot danger is to ensure rapid drying, generally by lifting coverings and floor boards, but also by removing skirtings from saturated walls and perhaps exposing concealed timbers such as lintels.

Some plumbing leaks do not cause dampness that is readily observable, but damage can still be very serious. For example, service and central heating pipes are often laid in notches in joists immediately below the floor boards, and sometimes copper pipes are accidentally penetrated by the floor-board nails. At first the nail will probably be a tight fit in the hole and there will be no significant leakage, but gradually the steel nail will corrode, as described in section 4.4, and this corrosion will eventually result in an increasing leak which will usually emerge at sufficient pressure to spray the surrounding floor boards and joists with a fine mist of water, eventually forming a small area of completely saturated wood. The area around the leak is often too wet for the development of the normal wood destroying fungi, but a moisture content gradient develops between this very wet area and the surrounding dry area, often supporting concentric rings of different fungi, each of them appropriate to the moisture content of the area in which they develop. In this case the fungal zoning will give a clear indication of the source of the problem, even if the leak has been repaired, its original position being indicated perhaps by an additional new connector in the pipe, but other types of zoning can also be important. For example, severe frost may fracture the rising high-pressure supply to a header tank in a roof space, subsequent thawing resulting in severe flooding due to the high pressure of the supply, the water spreading downwards through the building in a basically conical pattern. Insurance policies do not generally cover the cost of repairing such a leak, but they usually cover consequential damage through flooding. Often an insurance company will arrange for rapid replacement of carpets and repairs to decoration in order to minimise any claim for loss of use of the accommodation, but such rapid repairs generally ignore the danger of Dry rot development in poorly ventilated wood components. Dry rot therefore often subsequently develops, typically 6 to 24 months after the original flood, but insurance companies then often deny liability, and it may be necessary to relate the initial zones of infection to the extent of the original flooding.

Although the risk of wood decay is certainly the most serious consequence of leaks and floods, other materials may also be affected. Internally, wallpaper may become stained or detached, and gypsum plaster will deteriorate and will require replacement. Carpets can often be reused provided they are dried thoroughly and promptly, although they may require replacement if they are affected by staining or distortion such as shrinkage.

Drying buildings and their contents which have been affected by flooding can present problems, either through the large quantities of water involved or because the water is relatively inaccessible. There are several separate processes involved in drying. The first involves evaporation of water from material, and this can be divided into stages. The superficial surface moisture can be lost comparatively rapidly but the deeper moisture is more difficult to evaporate. In wood the removal of free moisture from the pores is followed below fibre saturation point of about 28% by the removal of combined water from the cell walls. The loss of this combined water will lead to shrinkage, but the danger of fungal decay development remains until the water content is reduced below about 20%. Consideration should always be given to preservation treatment if wood has moisture contents in excess of about 18%; at this moisture content fungal decay cannot develop, but it can spread from adjacent affected wood.

Remedial preservation treatment often involves the application of organic solvent preservatives which will only penetrate reliably at moisture contents below about 18%. For these various reasons drying is necessary to reduce the moisture content below 18% but also to recondition the wood to the ambient levels that it will achieve in service, perhaps 12–15% in most structural timbers but 8–12% in centrally heated buildings.

In theory, heat is required at a rate of 0.694 kWh/litre to achieve water evaporation. If heat is not supplied evaporation causes a decrease in temperature, the reason why a damp building feels cold. If the structure is heated to encourage evaporation and drying, the consumption will exceed this theoretical value as the structure itself will be heated and some heat loss will occur. The rate of evaporation depends upon the temperature, which in turn depends upon the rate of supply of heat. The vapour pressure of water, or its tendency to evaporate increases with temperature and doubles if the temperature is raised from, for example, 5 °C to 16 °C, so that heating of a structure is essential if evaporation is to be encouraged, and if heating is not used the evaporation will be very slow. A further problem is the drying of deepest parts of the structural members; the rate slows as the evaporation surface recedes into the material so that the evaporating moisture must diffuse to the surface, and it is thus essential to expose all structural members as far as possible in order to encourage drying.

The next process is to remove the evaporating water. As water is released into the atmosphere the humidity increases and this inhibits further evaporation. The water-carrying capacity of air increases as the temperature increases, as explained in section 4.2 and Figure 4.1. However, the vapour pressure of moisture in air does not depend directly on relative humidity or temperature, but only on the humidity or moisture content of the air. The importance of these facts is that, if the air outside a building is cooler, its moisture carrying capacity will be much lower and, even if it is saturated, it is unlikely to have a moisture content as high as the warmer internal air, so that there will be a tendency for the internal vapour pressure to be higher than the external vapour pressure and thus a movement of moisture towards the exterior. The most efficient method for drying a building is therefore to permit free circulation of external air, even if the external conditions are cold and the air apparently saturated. If the external air temperature remains below freezing for a significant period the moisture carrying capacity of the air is dramatically reduced, so that ventilating air has a very low relative humidity and can achieve very rapid drying. An important point to consider is the energy required for evaporation to occur from the structural material; whilst this can be provided by warming incoming air, perhaps by passing it through propane blower heaters, it is usually far more efficient to arrange for night-time heating to be alternated with day-time ventilation.

Dehumidifiers are often suggested as a means for reducing the relative humidity of the atmosphere and thus for drying structural materials. Hygroscopic absorbents have been used on a small scale and are, for instance, used in double glazing to ensure that the air gap remains dry, but mechanical dehumidifiers are essential if structural drying is to be attempted. Mechanical dehumidifiers usually involve a circulation fan which passes air over a low temperature condenser, the low temperature increasing the relative humidity and causing moisture to condense as dew which is then collected. Such dehumidifiers must operate in an unventilated space so that they can control the relative humidity of the air, but they also cool the circulating air and eventually evaporation from the structural material will cease through this cooling. This problem is usually corrected by heating the air leaving the dehumidifier, but the energy required for direct evaporation of water from structural materials is enormous, as previously indicated; dehumidifiers operated in an unventilated space are therefore extremely inefficient in relation to the volume of water that can be recovered; ventilation from the exterior is always most efficient in terms of the volume of water removed by drying and, particularly, the cost of drying.

If external materials become saturated with water through leaks or flooding, they become particularly susceptible to the frost damage described in section 4.3. It is therefore common to see an apparently sound masonry wall with a vertical strip of deterioration associated with, for example, a faulty downpipe or gutter joint, or a horizontal band of damage at the base of the wall associated with water splashing from overflowing gutters or passing traffic. Similar frost damage can also occur in free-standing walls and parapets if they are not provided with proper coping protection. Copings should always be designed to discharge rainfall clear of the wall by means of a projection provided with a drip, but problems may still arise from water penetration through fractured coping joints. Copings are subject to extreme temperature and moisture content fluctuations which cause movement stresses and the inevitable development of joint fractures; there should therefore always be a continuous damp-proof course beneath a coping to prevent moisture from penetrating through joint fractures into the masonry beneath. An important and often ignored requirement of this damp-proof course is that it should project clear of the masonry surface, as otherwise water retained on top of the course will run down the wall surface and be absorbed into the wall beneath the course. A very dramatic example of failure due to an inadequate coping damp-proof course occurred in Scotland some years ago. The walls of the buildings concerned consisted of cavity brickwork, rendered externally and projecting above the roof level as parapets. The felt roof covering was turned up the parapets and continued as a horizontal damp-proof course beneath the copings. Before completion of the building it was observed that there were stain marks on the harling or rendering beneath each coping joint, suggesting that the joint had fractured, and it was decided to rake out the joints and to seal them with a mastic pointing. A few years later, after a period of severe frost, it was observed that patches of render had spalled away beneath each coping joint. Investigation showed that the roofing felt damp-proof course had been trimmed flush with the outer brickwork surface before the render had been applied, so that water penetrating through the fractured coping joints was able to accumulate in the render beneath causing the observed staining and frost spalling damage. This parapet detail is not unusual but damage will not occur if the damp-proof course beneath the copings projects beyond the render surface.

4.7 Condensation

Air and moisture content relationships are discussed in detail in section 4.2 and presented in the psychometric diagram Figure 4.1. The risk of condensation depends simply on the moisture content of air and the temperatures that it will encounter; condensation occurs when air encounters temperatures below its dew point, and the dew point depends only on the moisture content of the air. The determination of the dew point for accommodation air is described in section 4.2, in which it is recommended that the ambient relative humidity of the accommodation air should be determined by measuring the moisture content of painted wood components isolated from any obvious source of penetrating, rising or condensing dampness, such as an architrave over an opening in an internal partition wall. The wood moisture content or ambient relative humidity determined in this way enables the dew point to be determined through an estimate of the average temperature of the accommodation air; it is suggested that 15°C is appropriate for living rooms in winter and 17°C in summer, with 12.5°C in winter and 15°C in summer for bedrooms. Dew points determined in this way must then be related to the temperatures that may be encountered. The most common form of condensation is dew on the cold internal surfaces of glazing, window frames or even walls, but Figure 4.3 enables the internal surface temperatures to be deduced from the thermal resistance R or thermal transmittance U values for a structural component in conjunction with internal and external temperatures.

Obviously dew is most likely to develop on the coolest surfaces, that is the internal surfaces of external wall components with the lowest thermal resistance R or highest thermal transmittance U values, but there are other factors which will affect the situation. Thermal bridging occurs when a structural element contains a feature which conducts the heat more readily. Local damp patches are generally due to thermal bridging, the dampness occurring only in the area of the conducting feature or bridge, such as where otherwise insulating external walls are bridged by dense concrete lintels, beams or posts. Even metal ties in cavity walls can cause thermal bridging and local damp patches on external walls, although it must be appreciated that accumulations of mortar slovens on ties can conduct moisture between the skins as explained in section 4.8. If external conditions are reasonably cold, thermal bridges can be identified by the very low temperatures of the areas affected by dampness, although it must be recognised that all isolated damp patches are cooler by up to about 1.5°C due to evaporation of moisture, so that it is sometimes easier to distinguish between condensation and penetrating dampness simply by observing whether it is most severe in cold dry weather or when the wall concerned is absorbing continuous rainfall.

Condensation on window glazing and frames is most severe when the windows are covered by curtains, even thin net curtains severely increasing the condensation. The explanation for this observation is that a curtain and, particularly, an air space between it and a window represents significant insulation; the inner surface of a window becomes cooler when the curtains are drawn, although the curtains do not significantly obstruct the diffusion of humid air from the accommodation towards the cold window surface. Exactly the same condensation effect can occur in cupboards built against external walls; mould growth on clothing through condensation is a common problem in wardrobes built against external walls and even loose furniture placed close to external walls can suffer deterioration problems through condensation. This effect is, in fact, a form of interstitial condensation, that is condensation occurring within a structural element. For example, the thermal calculations for a tile-clad timber frame wall are given in section 3.2 and are illustrated in Figure 3.3. If measurements of internal joinery have shown moisture contents of, for example, 17% (equivalent to an ambient air relative humidity of about 77%) and the average temperature for the room is considered to be about 15 °C, it is apparent from Figure 4.1 that the dew point for the air is about 11 °C. It is also apparent from Figure 3.3 that, when the internal temperature is 20 °C and the external temperature is 0 °C, this dew point occurs in the glass fibre insulation quilt, and there is therefore a serious danger of interstitial condensation causing decay of wood components on the outer side of this position, such as the sheathing plywood and the outer parts of the stud frames. This danger is normally minimised by introducing a polythene vapour barrier beneath the internal plasterboard lining and also ensuring that any moisture accumulating within the sheathing plywood can readily disperse to the exterior through a breather paper covering and the tiling. Where moisture susceptible materials are involved it is essential to assess the interstitial condensation risk of a structural element and to take appropriate precautions to minimise the condensation risk, both by the use of internal vapour barriers and external ventilation and also by using appropriate preservation treatment. For example, interstitial condensation can occur within wood window frames and is the most frequent cause of fungal decay damage. The traditional precaution is to construct the frames from naturally durable wood, but in recent years preservation treatments of non-durable woods have been extensively adopted as a realistic alternative; the problem can also be avoided by using the same technique as for wood frame structures, that is the use of an impermeable vapour barrier paint on the interior surfaces and a permeable breather paint on the exterior surfaces. The same principles must be applied to roof structures, particularly to flat roofs in which interstitial condensation problems are often encountered, as explained in section 13.1.

Dew formation on cold surfaces of structural elements is not the only form of condensation. Dew develops when humid air contacts a surface below its dew point, so that the danger of dew development is increased if the humidity of the air is increased or the temperatures of surfaces are reduced. Obviously dew development is therefore most likely in areas with high atmospheric humidities, such as bathrooms, kitchens and laundries, but humid air from these areas can diffuse into adjacent areas, such as corridors and larders, causing dew condensation and often persistent mould growth on affected surfaces. Even warm air diffusing from a poorly ventilated living room into a cooler corridor or bedroom may cause condensation and mould problems in this way. However, where virtually saturated air is involved, slight cooling will cause mist formation within the air itself, even without contact with cold surfaces. This is most obvious when, for example, a bathroom door is opened and the air flowing out can be seen as a cloud of mist.

In slow combustion coke and anthracite boilers the air flow is restricted to control the rate of combustion and the flow of air in the flue is thus extremely limited at times. The high humidity combustion gases pass slowly up the flue, becoming cooler and condensing, particularly at stack level. This condensation moisture causes damage to stone and mortar through absorption of acid gases, and these reactions often result in the formation of hygroscopic salts which migrate in the dampness to the plastered and decorated surfaces within the accommodation. Dampness and staining is frequently observed on chimney breasts, particularly those containing flues from slow burning solid fuel boilers. Modern gas and oil boilers are equally susceptible to condensation problems, mainly because of the low temperature of the flue gases. In all cases flue condensation is best avoided by increasing air flow; most modern boilers are fitted with open flues with free air flow, the boiler outlet being introduced into the side of the flue, an arrangement that also isolates the boiler from pressure fluctuations in the flue during windy weather. If open flues are not used it is necessary to line the flue both to avoid condensation damage and to provide a container to catch the condensation water dripping from the bottom of the flue!

Ventilation is an important feature in the avoidance of condensation dampness, but the most important point to appreciate is that warm air in a building invariably has a higher humidity or moisture content than cooler air externally; the external air may appear to be damper because it has a very high relative humidity, but this is because of its lower temperature and not because it contains more moisture. The vapour pressure of the moisture in air depends only on the humidity or moisture content, so that the vapour pressure of the more humid air within a warm building will be higher than the vapour pressure of even saturated external cool air. Ventilation must therefore be provided which will allow moisture in the warm internal air to diffuse to the exterior through the influence of the difference in vapour pressure; because of this vapour pressure difference excessive ventilation is not necessary. A permeable structural material, such as a brickwork wall, will actually allow sufficient diffusion to avoid most condensation problems, even if windows and doors remain tightly closed, but obviously problems occur in parts of the accommodation subject to exceptional humidity, such as bathrooms, utility rooms and kitchens, where more generous ventilation is essential. In kitchens and utility rooms the presence of a normal flue boiler can be very helpful in inducing ventilation, although there must be adequate ventilation from the exterior to provide necessary combustion air. However, since modern balanced flue boilers have no connection with the interior accommodation, as both their air inlets and flue outlets are external, they cannot function in this way, and alternative provision for ventilation is essential; it is this change in boiler technology that is the primary cause of condensation in some modern buildings. The usual solution is to install extractor fans in bathrooms, utility rooms and kitchens where humidity is generated, but extractors tend to cause excessive cooling, as explained in more detail in section 3.3. It is probable that future buildings will incorporate special ventilators in these rooms which will be fitted with membranes allowing water vapour to vent to the exterior without unnecessary loss of heat, although membrane systems of this type are only available at present in some dehumidifier systems such as the Munter Rotaire systems.

4.8 Penetrating dampness

Water penetration through walls is generally avoided through precautions in design and construction. Overhanging eaves considerably reduce rain penetration into wall surfaces, except during windy weather, and even where flush eaves are employed, gutters and downpipes are provided to ensure that walls are not exposed to roof water discharge. However, these precautions should not be necessary as walls should be capable of resisting rain penetration in any case. In solid walls very thick construction is used so that absorbed penetrating rainwater can be accommodated without becoming apparent at the internal surface, the accumulated water later dispersing by evaporation. In cavity walls complete penetration and saturation of the external skin is expected, but precautions are taken to prevent absorption of this water into the internal skin. In both cases defects can occur and internal dampness problems are often encountered.

In solid walls penetrating dampness may result from insufficient absorptive capacity, penetration occurring because the wall is too thin, either throughout the building or at local thin points, such as window reveals. Alternatively, penetration results from excessive permeability. With carbonaceous stones, that is limestones or sandstones with a carbonate cementing matrix, atmospheric acids will cause slow erosion and a progressive increase in permeability, as described in sections 4.4 and 7.2. In many cases dampness follows the pattern of the masonry, developing in either the mortar or the stone, whichever is more permeable. This effect can be seen most clearly if the internal surface is finished in limewash, the custom in many old churches in some parts of the British Isles. Construction in impermeable granite presents a particularly interesting example. The impermeability of the stone would seem to be a good protection against penetrating dampness, but the same amount of rainwater will be incident on an impermeable granite wall as on any other wall; all the water flowing down the wall will be absorbed into the porous mortar which, with its limited absorptive capacity, will quickly become saturated throughout the thickness of the wall, although the mortar might have perfectly adequate capacity if used in combination with a stone of average porosity which could absorb some of the water. In fact, many granite walls are not constructed in this way, but comprise two separate skins with a rubble core. Penetrating water can drain through the core, emerging at the interior at some distance from the original source, making it difficult to positively identify a defect such as an area of faulty pointing.

Penetrating dampness resulting from inadequate absorptive capacity tends to result in uniform internal dampness, except where the upper parts of walls are drier as they are protected from rainfall by overhanging eaves. However, rainwater absorbed into a wall will tend, at least in durable macroporous materials, to drain towards the base. If a damp-proof course is provided it is likely that this water will accumulate on top, giving an appearance very similar to rising dampness, although often the wall is dry underneath the damp-proof course. Whilst this accumulating dampness may appear similar to rising dampness, treatment is obviously entirely different and it is thus essential to ensure that dampness of this type is correctly diagnosed. Similarly, if dampness is concentrated at the top of a wall it must be suspected that there is a roof or adjacent gutter defect, and heavy flow from a defective gutter or downpipe may cause a vertical band of dampness. Parapet and valley gutters, as well as flat roofs, frequently drain into hoppers which lead to downpipes. In some large buildings the hoppers may be fitted with spouts at a higher level through which water can discharge if the hopper should become blocked. Dampness in many buildings can be attributed to the failure to keep hoppers free from accumulating leaves, moss, lichen and pieces of stone, so that the entire roof discharge eventually occurs through the overflow spout, or simply overflows down the wall if spouts are absent.

Some years ago extensive renovation works at a Scottish castle were followed by the development of dampness at the junctions between walls and ceilings, immediately below every parapet, a problem that had never occurred previously. Investigation showed that the parapet copings had a slight fall towards the roof, but their edges did not project beyond the parapet walls so that water was discharging down the roof side of the parapet, being absorbed and percolating downwards to the accommodation beneath. The problem had not occurred previously because the roof asphalt had been continued up the inner faces of the parapets and across the copings, a detail that had been introduced by the well-known architect Lorimer when the castle was repaired and extensively reconstructed in about 1908 following a fire. This damp-proof membrane feature was considered unacceptable by the architect preparing the scheme for the recent renovations and it was removed, but unfortunately the architect failed to appreciate the essential damp-proofing function of the asphalt covering. The dampness was cured by reconstructing the parapets with damp-proof courses coupled with flashings at the top of the asphalt upturns.

In theory the use of cavity walls should completely prevent penetrating dampness, but unfortunately practice is not as perfect as theory! Dampness at the top of walls can arise through roof, parapet and gutter defects, as for solid walls, but even direct penetration can occur, spreading across the cavity through mortar droppings or slovens which accumulated on wall ties during construction, causing small patches of dampness internally. Sometimes this problem develops only when cavity fill insulation is introduced, the original ventilation of the cavity being sufficient to prevent significant water penetration across the ties but the cavity fill destroying this ventilation and causing the dampness to become apparent as patches on the interior surfaces of the walls. Cavities should preferably extend to well below internal damp-proof course level or, alternatively, if the damp-proof courses are continuous, they should be stepped so that the level in the external skin is lower than in the internal skin. Occasionally the situation is reversed, either through carelessness or ignorance, and penetrating dampness accumulating on the damp-proof course at the base of the external skin may flow into the internal skin; in one example this defect occurred in all the houses throughout a large building development!

In another house sulphates in the bricks reacted with the cement in the mortar of the external skin, initially causing expansion and distortion. This prompted some reconstruction of the brickwork in some areas, but where the original brickwork had been retained the mortar eventually became very friable; sulphate attack in brickwork is described in section 7.3. One day there was a severe thunderstorm, the thunder shocks causing the friable mortar to run from the external skin into the cavities, accumulating on the trays over openings and providing bridges through which the heavy rainfall penetrated from the external into the internal skin, damaging the interior decorations!

It has become normal practice in recent years to restrict cavity ventilation to improve thermal insulation. In fact, cavity ventilation has a very important function in evacuating humid air; if vents are omitted dampness penetrating through the external skin will increase the humidity of the cavity air. If the wall cavity is continuous with the roof space, and the roof space is similarly unventilated, condensation will eventually occur if the roof covering is impermeable, water accumulating in the supporting boarding of a wood roof deck, perhaps resulting in the development of fungal decay within the boarding and the supporting joists, and in severe cases causing dampness staining through condensation dripping onto the ceilings beneath. Cavity fill insulation reduces heat loss by restricting the convection circulation that transfers heat across the cavity from the inner to the outer skin, but fill materials must be chosen with care. Non-wettable pelletised materials, such as expanded polystyrene, can combine excellent thermal insulation with freedom from disadvantages, but some mineral fibre materials can conduct moisture across cavities; the mineral fibre should be treated with a water repellent but there have been many instances of cavity fill without water repellency, either because of a fault in manufacture or because the water repellent has been destroyed by biodeterioration. Even foams formed *in-situ* can have considerable disadvantages; wettable injected foams can conduct moisture across the cavity, and if they are incorrectly formulated they can actually collapse to release moisture and cause dampness. There are also two forms of dampness development which are common to all forms of cavity fill insulation. If mortar slovens were permitted to accumulate on cavity ties during construction, they will tend to conduct moisture across the cavity from the outer skin to the inner skin, but in a normal ventilated cavity the evaporation from these sloven bridges is usually sufficient to prevent any dampness becoming apparent within the accommodation. However, all forms of cavity fill restrict ventilation and increase the likelihood of dampness developing in isolated patches through sloven bridges. In addition, the vibration caused by drilling into the outer skin in preparation for cavity filling often loosens considerable debris from the inner face of the outer leaf which then accumulates at the bottom of the cavity, perhaps bridging trays over openings and damp-proof courses, causing dampness to develop in the accommodation; this problem is particularly severe in seaside properties in which sea spray absorbed into the outer leaf can cause crystallisation damage to the inner surface which can then be loosened by the drilling vibration.

4.9 Rising dampness

Water absorbed by capillarity into the base of a wall from the supporting ground is generally described as rising dampness. Uniform dampness confined to the base of both external and internal partition walls can usually be attributed to the lack of effective damp-proof course. Obviously a damp-proof course may be discontinuous, perhaps around hearths or where old doorways have been blocked, and localised dampness may therefore occur. Rising dampness is not confined to walls, but also occurs in solid floors if a proper damp-proof membrane has not been provided. In some cases damp-proof courses are present in walls and damp-proof membranes in solid floors, yet rising dampness still occurs through discontinuity or bridgings.

Damp-proof courses were often positioned above the level of solid floors in Victorian houses, presumably because it was considered that solid floors were regularly washed and the walls therefore required protection against the washing water as well as dampness rising from the ground. Usually a batten was fixed to the walls immediately above the damp-proof course providing an edge to the plasterwork above, effectively preventing the plaster from bridging the damp-proof course and also providing a fixing for the skirting which was thus isolated from the wall below damp-proof level by an air gap; this form of construction was the reason why Victorian builders used such deep skirtings. Whilst this system was perfectly logical, the door frames extended below the damp-proof course and absorbed moisture, often becoming affected by Wet or Dry rot which then extended into the adjacent skirtings. Scarf repairs to door posts and replacement of skirtings were simply followed by further fungal decay, and eventually the skirtings were abandoned in many cases and replaced with a dense rendering. The effect of the rendering was simply to bridge the damp-proof course, causing deterioration of the plaster and decorations above.

In modern construction, bridging can occur in many different ways, the most common perhaps being the failure to link the damp-proof membrane in a solid floor properly with the damp-proof course in the adjacent walls. Generally the damp-proof membrane should be turned up at the edges, led up the wall and tucked into the damp-proof course during construction, but this is actually a difficult detail to achieve; adhesive damp-proof membrane such as Bituthene is often helpful in linking the two membranes, or in carrying out remedial work. Similar bridging can occur externally if walls are rendered, or even if the damp-proof course joint is pointed for aesthetic reasons.

An important point is that penetrating water in external solid walls may accumulate on an existing damp-proof course, and this accumulation may be incorrectly diagnosed as rising dampness. The insertion of a new damp-proof course will not remedy this problem, the most appropriate treatment being application of a surface water repellent to reduce penetrating dampness. Diagnosis of the source of dampness is therefore particularly important.

4.10 Hygroscopic salts

Hygroscopic salts are generally introduced into walls in rising dampness and accumulate at surfaces where evaporation occurs. Less commonly they are introduced in penetrating dampness, usually from sea spray or industrial pollution, although in polluted urban atmospheres penetrating dampness contains acids which react with the structural materials, as described in section 4.4, to produce salts which may be hygroscopic. A precisely similar situation occurs through the acids in flue gases from coal or oil combustion, particularly where condensation occurs in flues, as explained in section 4.7. In all these cases salt deposits occur at evaporating surfaces, often on decorated surfaces in the accommodation, and when these salts are hygroscopic they will absorb moisture from the atmosphere, particularly in humid conditions, and produce dampness even in the absence of a source of moisture. As the salt deposits absorb moisture they obviously restrict evaporation, and when these problems occur in association with rising dampness the salt deposits force the dampness to rise progressively further up walls. The diagnosis of dampness caused by hygroscopic salts has been described in section 4.2.

Patches of hygroscopicity can be very difficult to explain and it may be necessary to enquire into the detailed history of a building. The generation of hygroscopic salts within the construction materials through flue gases has already been described, but it must be appreciated that the salt deposits will persist, even if the flues are removed. In older buildings patches of hygroscopic dampness can sometimes be attributed to combustion gases from old coppers or bread ovens which have been removed, and in other cases salt deposits can be attributed to storage of coke, perhaps on the other side of a wall; coke frequently contains hygroscopic salts and, as it is sometimes wet when delivered, these are able to migrate into the floor and walls of a fuel store. In other cases salt deposits are caused by bacon curing, or storage of salt for use in water softeners or treatment of icy roads and paths. In many cases chemical analysis is an essential preliminary to any attempts to identify the sources of the hygroscopic salt deposits.

In all cases it is necessary to cure sources of dampness and remove salt deposits. Some damp-proofing contractors place considerable emphasis on the need to remove salts from within walls but, in fact, rising or penetrating dampness usually contains only minute concentrations of salts; significant accumulations occur only at evaporation surfaces where they can be easily removed by stripping and replacing plaster, or by a process of alternate water spray and brushing off salt efflorescence if bare brickwork or stone masonry is involved. Various products are promoted as salt inhibitors, retarders or neutralisers but no such chemicals actually exist! Salt migration through masonry and plaster can only occur in solution, so that prevention of moisture movement will also prevent salt accumulations. The essential feature of a salt inhibitor, retarder or neutraliser is therefore a waterproof barrier that will prevent the migration of salt solution into plaster from the supporting masonry, but some products offered for this purpose can cause serious problems. Calcium chloride is sometimes marketed as a waterproofing additive for cement mixes, such as renders and undercoat plasters, but it is actually a cement cure promoter which is usually used as an accelerator, any waterproofing action arising only through curing promotion which tends to result in a denser material. If it is used as a salt inhibitor it is usually completely ineffective, but it can cause severe corrosion of electrical and plumbing fittings.

4.11 Damp-proofing treatment

Walls are traditionally protected from rainfall by overhanging eaves fitted with gutters and downpipes. However, rain is often blown onto wall surfaces, the danger of penetrating dampness depending on the degree of exposure, that is the total rainfall in relation to wind velocity. On a normal porous wall the surface is wetted in this way during rainfall and the water is then absorbed by capillarity; the force drawing the water into the wall depends on the surface tension and the contact angle for water in relation to the pore size. With smaller pores the capillary force is greater, but in all normal porous walls the force is sufficient to ensure that water is absorbed into the wall rather than flowing down the external surface. The total porosity of the wall in relation to its thickness will determine whether the wall can absorb the rainfall without it becoming apparent at the interior surface. In addition, it is found that a macroporous wall surface will allow water to evaporate more rapidly to the exterior than a microporous surface, and as the latter is likely to retain moisture for a longer period it is also more likely to permit rain penetration to the interior and to result in frost damage at the exterior surface. If a wall consists of relatively impermeable blocks of stone the process is slightly different; the water flows down the impermeable blocks and is absorbed into the mortar courses and, as the mortar represents only a fraction of the total wall volume with a comparatively low capacity for water absorption, penetration to the interior is likely at the joints.

One method for reducing water penetration through walls is to increase their absorptive capacity, usually by the use of a thick porous render, a process that also significantly improves thermal insulation value. Unfortunately this process is not generally understood; often thin dense or waterproof renderings are employed which are not usually sufficiently flexible to tolerate the seasonal thermal and moisture content changes in the structure so that eventually cracks develop. Water flowing down the wall is readily absorbed into these cracks, yet the remaining render obstructs evaporation so that the final result tends to be progressively accumulating moisture and the development of increasingly severe dampness internally. Remedial works to reduce this dampness often include the use of cement slurry or paint, or water-proof bitumen, oil paint or plastic coatings, but all these systems are unsuitable as they introduce severe interstitial condensation dangers; a permeable structure must never be sealed at the cold external surface as there is then a danger that interstitial humidity will condense immediately beneath the water-proof layer, followed by severe frost spalling in cold weather.

Rendering and coating systems suffer from the disadvantage that they completely change the appearance of walls, but water repellent treatments provide an invisible method for reducing water penetration into porous walls without the condensation and spalling dangers associated with impermeable treatments. Water repellents act by lining the pores so that the angle of contact is reversed and the capillary forces then repel rather than absorb the water. When first applied water repellent treatments cause rain to globulate on the surface so that it falls away rather than being absorbed but, with most water repellents, this globulation or 'duck's back' effect is soon lost and rain is able to wet the surface. However, this wetting is only superficial and, if the water repellent treatment has been applied sufficiently generously, the pores remain water repellent to an appreciable depth so that water absorption is still prevented. Water repellent treatment is inexpensive as it involves only a rapid generous spray treatment, but the treated wall has a number of advantages in addition to the simple control of penetrating dampness. A dry wall has better thermal insulation properties, as explained in section 3.2, but it also remains more uniform in colour during wet weather. It is, of course, possible for water to accumulate behind a water repellent treatment, either through interstitial condensation or through wind pressure overcoming the water repellency on macroporous materials used in extremely exposed situations, but true water repellents do not seal the pores, and any water introduced in these ways is able to disperse by evaporation to the exterior. However, it must be firmly emphasised that only silicone resin (siloxane) water repellents should be used, as they are the only compounds that have sufficient durability when applied as the very thin pore linings that are necessary if permeability is to be maintained so that trapped moisture can disperse. Other water repellents are promoted for the treatment of external wall surfaces against penetrating dampness, particularly formulations based on wax, vinyl or aluminium compounds; however, if they are applied at sufficient concentrations to give resistance to water penetration and reasonable durability, they cause darkening of porous materials and they also restrict permeability and reduce water evaporation to a much greater extent than equivalent silicone resin treatments.

British Standard specification BS 6477:1984 *Water repellents for masonry surfaces* classifies water repellents according to their suitability for use on four types of substate: group 1 is predominantly siliceous including clay bricks, sandstones and mature cement-based materials, group 2 is predominantly calcareous including natural limestones, group 3 is fresh cementitious and other alkaline materials, and group 4 is calcium silicate brickwork, commonly known as sand-lime or flint-lime brickwork. Water repellents are tested for early water repellency, absorption of water, evaporation of water, and durability. Unfortunately the test requirements are very easy to meet. A very superficial water repellent treatment is sufficient to achieve the early water repellency requirement. The water absorption test over 7 days allows the test specimens to absorb up to 10% of the water that will be absorbed through an unprotected surface. The evaporation of water test requires the evaporation from the treated specimens to be only more than 10% of that of untreated specimens. The durability test requires only that these criteria will be met after the specimens have been weathered for 24 months. These requirements do not therefore take account of service experience with different types of water repellent and represent really a method for selecting formulations that might be suitable rather than formulations that are proven to be reliable in service. In contrast an earlier British Standard BS 3826:1969 *Specification for silicone-based water repellents for masonry* was restricted to silicone water repellents, as it was considered that they were the only formulations that had been shown in service to have adequate durability. BS 3826 defines the three principle types of silicone water repellents that are available. Class A is a normal silicone resin treatment in an organic solvent, such as white spirit or kerosene, but it gives unreliable results on carbonaceous substrates such as limestones; the definition in BS 6477 of limestones as calcareous substrates is misleading as it suggests that differences occur through calcium content, whereas it is actually the carbonate content that causes unreliability with certain silicone resins. Class B is a more sophisticated resin which will give reliable results on all normal porous substrates; most organic solvent silicone resin treatments now conform to both Class A and B. Class C is limited to siliconates; these are water soluble formulations which cannot be generally recommended as they have a comparatively short effective life.

Whilst all British Standard water repellent systems are required to retain sufficient permeability on normal substrates to permit trapped water to disperse, it must be appreciated that dampness rising behind the treatment may introduce salt deposits which can cause the treated layer to spall away, so that surface water repellents should never be used in circumstances where walls are subject to rising dampness. In addition, water repellents should never be applied to walls that are covered with moss, lichen or algal growth or walls with defective pointing or excessively friable stone. All these problems must be remedied before water repellent treatment; section 7.7 describes suitable biocidal treatments.

Water repellent treatments can achieve particularly dramatic results in some circumstances. Walls constructed from impermeable stone present particular problems, as previously explained, as the entire absorbed rainfall is concentrated at the mortar joints. In many cases these joints are subject to cracking due to seasonal thermal and moisture content changes, perhaps allowing considerable rainwater absorption and development of severe dampness at the interior. These problems are not confined to common impermeable stone, such as dense quarried granite, but apply equally to pebble and knapped flint walls. With these forms of construction the joints are often too thin to permit repointing, but spraying the wall with a water repellent formulation can often achieve dramatic control of rain penetration. The spray must be applied generously so that it flows down the wall surface and is absorbed into all cracks, crannies and porous areas that will normally absorb rain-water; indeed, this is the normal method for applying surface water repellents. A coarse low-pressure spray should be used, the spray lance being moved horizontally across the surface to produce a run-down of perhaps 150 mm (6"), the next pass being across the run-down to produce a further run-down, and so on so that a continuous curtain of water repellent fluid flows down the wall until the entire surface has been treated in broad vertical stripes. Usually the solvents in the treatment will persist for a sufficiently long time for treated areas to be readily identified, although some products are available containing fugitive dyes which stain the wall at the time of treatment, the colour then gradually disappearing over several weeks.

Where solid walls are used, such as in domestic garages and in industrial buildings, serious penetrating dampness can arise if the construction materials are excessively permeable, some cast stone blocks being very troublesome in this respect. An inexpensive water repellent treatment can usually completely prevent rain penetration, but it will also substantially improve the thermal properties of the wall; a dry wall has much greater thermal resistance than a damp wall, and evaporation at the external surface of the damp wall can considerably reduce wall temperature and further increase heat loss. Perhaps surprisingly, many water repellent treatments are applied to new cavity walls. Outer leaves of excessively permeable natural or cast stone can permit excessive rain penetration into the cavity which may be sufficient to overload normal precautions, such as trays and vertical damp-proof courses at cavity closures, particularly if weep-holes are omitted at trays. Cavity bridging through mortar slovens or droppings which have accumulated on wall ties during construction usually leads to isolated damp patches internally opposite each tie bridge, particularly in conjunction with cavity fill insulation as previously explained in section 4.8, but this is another problem that can be remedied by a simple external water repellent treatment. Mortar droppings can also accumulate at the bottom of cavities and, if these are poorly designed, bridging can occur which may give the appearance of rising dampness, whereas rain penetration may be responsible. Accumulations at the bottom of cavities can occur through progressive masonry deterioration, usually at the internal surface of external leaves exposed to sea spray and particularly following drilling to insert cavity fill insulation which dislodges this debris. This is not, however, the only form of bridging that can develop following cavity insulation. Mineral fibre and similar insulation must be treated with a water repellent during manufacture to prevent cavity bridging, but it is often found that the treatment is ineffective, either because it was not properly applied or because it has deteriorated within the wall perhaps through bacterial attack; severe dampness can develop as a result. Where wall ties are bridged by accumulations of mortar slovens, cavity ventilation may be sufficient to prevent dampness appearing at the interior, but cavity fill destroys this ventilation and classical tie bridging dampness may develop as a result.

Problems with surface water repellent treatment are generally associated with errors in diagnosis of the dampness or carelessness in application of the treatment. All water repellents are sensitive to the moisture content and temperature of the surface; the surface must be 'white' dry, and extremely cold or hot surface conditions must be avoided. The organic solvent silicone resin treatments are usually most tolerant, but polyoxoaluminium stearate treatments are least tolerant as they gel by reaction with water and application to a damp surface results in the formation of an unsightly greasy coating.

Many 'specialist' damp-proofing contractors advertise only treatments against rising dampness by installing damp-proof courses, yet most apparent rising dampness cannot be remedied in this way. If apparent rising dampness is confined to solid external walls it must be suspected that it is penetrating dampness percolating downwards and accumulating on an existing damp-proof course, or rising dampness through bridging of the damp-proof course, perhaps due to the presence of a rendered plinth or even a line of pointing bridging the damp-proof course externally. Indeed, bridging of a damp-proof course is the commonest cause of true rising dampness and is correctly remedied by removing the bridge rather than by attempting to insert a new course.

There is, of course, a danger of rising dampness in all parts of a building in capillary contact with the ground, particularly the external and partition walls, the sleeper walls supporting plates and joists, and solid floors. In new construction an impermeable damp-proof membrane should be present throughout and, if there is no such membrane, the best remedy is always to insert one. It is relatively easy to introduce membranes under plates and joists, and even to incorporate a new membrane in a solid floor, perhaps using bitumen or an epoxy resin composition beneath thermoplastic tiles or wood strip flooring. Walls are more difficult and must be considered separately, but all damp-proof membrane systems must be continuous; the most common fault in new buildings is a failure to link damp-proof membranes in solid floors with damp-proof courses in adjacent walls.

With walls it is essential to check for the presence of an existing damp-proof course; if a course is present it is probably effective with rising dampness caused by bridging, and the correct remedy is the removal of the bridging rather than the installation of a new course which may well damage an existing course. If there is no damp-proof course, it is possible to insert one by carefully raking out mortar courses over a short run and inserting overlapping slates, metal sheet or bitumenised felt, as in a new structure, although systems are now available in which saws can be used to cut slots in walls to achieve the same result in a rather more efficient manner. In all such cases there is a danger that a building will settle; this can be very significant around door posts which must be trimmed if cracking of the structure is to be avoided. Another alternative is to run resin into the slot, a process that can avoid settlement if it is carried out carefully in short runs leaving intermediate sections which will support the structure whilst the resin cures and which can be separately treated later. Some resin systems involve slipping a plastic bag into the slot which is then filled with resin, a particularly neat but rather expensive system.

Whilst the introduction of a solid membrane in this way is definitely most reliable, it is unsuitable in some circumstances such as around fireplaces and in random rubble masonry walls. A chemical injection damp-proof course is much easier to use but results can be rather inconsistent. There are two basic techniques. The first involves diffusion from drillings. In the original process large diameter holes were drilled downwards at an angle of about 45° and were topped up continuously with water repellent until the desired absorption had been achieved. More recently, the topping up process has been replaced by bottles feeding automatically through tubes. The original processes used solutions of waxes in chlorinated solvents, such as trichloroethylene, but in recent years aqueous solutions of siliconates have been used. Drillings sometimes lead to cracks or open vertical joints, allowing treatment to drain away into the footings rather than diffusing into the adjacent walling materials to form a continuous water-repellent barrier, but in the Peter Cox transfusion process bottles are used to feed porous sponges which allow the treatment to flow by capillarity into contacting walling without draining to waste into cracks and cavities. However, it is difficult in relatively impermeable masonry to achieve adequate diffusion to produce a continuous water-repellent zone, whilst in very permeable masonry the flow of rising dampness may carry the water repellents further up the wall before curing occurs so that the damp-proof course may be discontinuous or formed too high within the wall.

An alternative treatment against rising dampness involves pressure injection of non-aqueous water-repellent solutions into tubes sealed into holes in the masonry. It is difficult to obtain the necessary distribution without wasting material through cracks and cavities; many operators prefer to use such processes only in brick walls, usually injecting into the bricks alone and hoping that sufficient diffusion will occur into the adjacent mortar, although other operators prefer to treat a horizontal mortar joint. Whilst all these injection damp-proof course processes can achieve a reasonable degree of reliability if used intelligently and installed conscientiously, they cannot be considered general remedies for rising dampness that can be employed by inexperienced staff; in some areas their use has fallen into disrepute as a result of the high level of failures. With pressure injection, low viscosity products must be used to achieve rapid distribution over a complete horizontal zone within the wall, but the treatments must then cure rapidly to fix them in position. In referring to low viscosity it is not sufficient to use a solvent of low viscosity, but it is equally necessary to use a polymer system with a relatively low molecular weight so that it does not suffer excessive filtration during the injection process; there is no point in achieving adequate distribution of the solvent if the water repellent has been filtered out on the surface of the injection hole! Generally, rapid curing silicone resin systems seem most effective, but in many relatively porous materials polyoxoaluminium compounds can give very reliable results. Many experienced contractors have an excellent reliability record for chemical injection damp-proof courses, attributing the relatively low number of failures that they suffer to difficulties in drilling some walls, particularly rubble walls constructed from very hard stone.

Electro-osmotic damp-proofing has been extensively promoted as a remedy for rising dampness, although in recent years it seems to have been recognised that the system is the least reliable of all the methods that are available; in investigations by the author since 1965 into damp-proofing failures, more electro-osmotic failures have been investigated than all other systems. The use of electro-osmosis for damp-proofing is based on the classical observation that water flow can be induced in a capillary when there is an electrical potential difference between each end. In 1807 Professor Reuss, working in Moscow, placed electrodes in wet quartz dust and demonstrated that a direct current would drive the water from the anode to the cathode. It was the two Ernst brothers in Switzerland who, in 1930, adapted this process as a remedy for dampness in buildings and applied for patents in various countries.

Electro-osmotic systems of this type, in which a current is passed through the wall, are known as 'active' systems. It was soon found that these systems possessed the serious disadvantage that a metal anode is destroyed by electrolysis. However, the Ernst brothers also observed that electrodes installed in damp walls possessed a potential relative to earth which, they concluded, might be inducing the rising dampness. They therefore suggested that the dampness could be prevented by connecting or shorting-out the electrodes, or connecting wall electrodes to earth, in order to remove this potential. 'Passive' electro-osmotic systems of this type have been installed in the British Isles since they were first developed by the Ernst brothers, but more recent passive systems involve more elaborate electrode systems which have been claimed to be more effective. Certainly, wall masonry generally possesses a positive potential relative to earth and, if this potential is truly responsible for capillary absorption of dampness, a simple short circuit might well induce drying. However, the short circuit might operate in reverse, giving the wall electrode system the same potential as earth and enabling the dampness to rise still further. Alternatively, the flow of water in the wall may be induced by the current flow between the zones of different potential; it can then be suggested that the electrode system provides a low resistance path which avoids current flow through the water in the wall so that rising dampness no longer occurs, but this theory ignores the fact that a circuit must be complete and that current will flow in the electrode only if water flows in the wall, a short circuit presumably increasing the dampness.

Certainly the case for passive electro-osmosis is unconvincing. It is said that the rising dampness is caused by the presence of electric charges and it is suggested that they are associated with the surface tension forces which induce capillarity. It is also said that these forces act in a consistent direction, water always moving towards the positive charged area; however, no explanation is offered for the fact that the potential is sometimes reversed relative to the direction of water flow, nor is there an explanation for the fact that these potentials are absent in laboratory experiments with clean porous materials and de-ionised water. However, the potentials develop immediately a soluble salt is introduced into the rising dampness, and it is clear that the potentials are, in fact, caused by an ionic diffusion gradient. Thus sodium sulphate in a wall will dissociate into positive and negative ions within the rising dampness. The negative sulphate ions are larger and tend to be obstructed as the dampness rises in the wall, so that the upper part of the wall possesses a higher concentration of positive sodium ions whilst the lower part possesses a higher concentration of negative sulphate ions. Reversal of potential is simply an indication that the relative sizes of the positive and negative ions have been reversed for the salt that is actually involved. Thus the potential within the wall is not causing the capillary absorption but is simply resulting from it, and the potentials that are found within walls cannot be used for the suppression of dampness.

Electro-osmotic damp-proofing systems have been investigated but there is no evidence that passive systems work or are theoretically capable of working; the position for active systems is less clear. Certainly Professor Reuss established that water flow can be induced between electrodes placed in wet quartz dust but, unfortunately, many translators describe the experiment as occurring in sand. Simple laboratory investigations are sufficient to demonstrate that electro-osmosis can only occur in very fine pored substrates. In coarser pored substrates, flow appears to be induced through movement of charged ions rather than through electron flow, so that the rate and direction of induced water movement tend to depend on both the nature of the substrate and the nature of any soluble salts that may be present. Thus passive systems can achieve no control over rising dampness but active systems may be able to induce reverse capillary flow in one way or another under suitable conditions, although the direction of flow will depend on the identity of the soluble salts that are present. Even if care is taken to ensure that electro-osmosis is used only when water movement can be induced in the required direction, problems can be encountered through destruction of the anodes by electrolysis, unless care is taken to use resistant anodes, such as carbon or platinised titanium.

When an active electro-osmotic system is operating it is usually observed that the current progressively falls. Contractors frequently claim that this is due to drying of the wall but most physicists and chemists will consider that it results from polarisation of the electrodes. If either explanation is correct, disconnection of the system for a sufficient period should result, on reconnection, in restoration of the original current, but it is actually found that this is rarely the case and a permanent reduction in current occurs until the flow eventually becomes insignificant after a period of a year or two. This diminution in the current flow is sometimes accompanied by a noticeable reduction in rising dampness, although this can only be observed clearly in the laboratory or in walls that are not affected by penetrating rainwater or other sources of dampness. The apparent explanation for this phenomenon is that ions within the wall, particularly calcium and aluminium, have been caused to migrate in the current field and have become redistributed, forming a densified zone between the electrodes, which obstructs both further current flow and rising dampness. Whilst this phenomenon has been observed in many walls in which active electro-osmotic systems have been installed, it is not clear that this effect is universal, and it appears to be most apparent in walls incorporating carbonaceous stones or lime mortars. A further and most important point is that this effect is observed only where sufficiently high current densities are employed. Current density depends basically on the applied potential divided by the distance between the electrodes, and if this effect is to be achieved it is essential to employ adequate potential whilst keeping the distance between the electrodes to a minimum. For this reason, this effect is seldom seen in systems which employ earth electrodes, but only in systems where both anodes and cathodes are inserted within the wall and to which relatively high potentials are applied; in the British Isles it is common to find potentials of less than 20 V, often as low as 6 V, used even with earth rod systems, whereas in Austria and Germany where damp-proofing with the Meikro carbon rod system has proved rather more effective, typical potential gradients of 1 V/cm of wall thickness are used between wall electrodes, potentials as high as 100 V or more being used for thick walls. Certainly, experiments by the author on actual sites in the British Isles have shown that active electro-osmotic systems can be effective if the applied potential is sufficient and the separation between the anode and cathode is minimised to enhance the potential gradient, provided that particular care is taken to embed the electrodes in the wall to ensure good electrical contact.

High capillarity tubes are another system that has been extensively promoted for damp-proofing. These tubes were originally developed about 65 years ago by the late Lt-Col B. C. G. Shore, a well-known innovator of conservation processes, particularly for use in natural stone masonry. He arranged for the tubes to be manufactured by Royal Doulton. In recent years other companies have promoted them as Doulton tubes without apparently being aware of their true origin. Knapen tubes perform a similar high capillarity function. The tubes are bedded in porous mortar in the walls affected by dampness, either horizontally or with a slight downward slope towards the outside of the wall. In theory, moisture within the wall migrates to the internal surface of the porous tube where it evaporates, the humid air flowing out of the tube and thus inducing a ventilation process which encourages drying; whilst it is normally expected that the humid air will be dense and flow downwards, it is actually less dense than dry air and flows upwards! Obviously, tubes must discharge to the exterior as they will otherwise serve only to increase the atmospheric humidity within the accommodation. Whilst properly manufactured and installed porous tubes can greatly increase the rate of evaporation from a damp wall, it must also be appreciated that the diffusion of dampness towards the tubes will transport salts which will be deposited on the evaporation surfaces where they will progressively reduce the evaporation rate, not particularly by clogging the pores but by their hygroscopic action which reduces evaporation. It is therefore necessary to replace tubes at regular intervals, usually about every 5 years, to maintain the evaporation rate. However, it must also be appreciated that porous tubes should never be used if an alternative damp-proofing process is realistic; they simply divert dampness without correcting the cause, and dampness can be controlled much more reliably by preventing rain penetration, rising dampness or condensation.

Reliable tanking is required when accommodation is below ground level. There are four essential requirements: the waterproofing must be completely effective even if subjected to hydrostatic pressure; it must bond reliably to the background to resist any hydrostatic pressure; it must bond to any internal finishing plaster; and dampness must not accumulate on the internal surfaces through condensation. Bonding to the background can be achieved most easily with cement-based materials, particularly if a wet cement slurry is applied to the background as a primer before applying a rendering or screed. Waterproofing additives, particularly stearates, can be very effective when incorporated in renders or screeds, although if hydrostatic pressure is expected it will be necessary to apply a special coating such as Vandex to the render undercoat, before applying a final coat of render and plaster finish. Condensation can be most easily avoided by providing thermal insulation between the waterproof system and the background, most simply by using a light-weight aggregate render system with the membrane on the warm internal surface. Another necessary precaution with tanking is to ensure the continuity of the membrane; failures often occur because there are gaps in the system at doorposts and partitions.

Where a tanking is provided in this way it must be linked with a horizontal damp-proof course through the external walls at about 150 mm above ground level. It is unusual for the tanking to be continued above this damp-proof course, yet there is no technical reason why internal tanking cannot be used to remedy rising or penetrating dampness; indeed, damp-proofing contractors often specify that water-proof plaster systems should be used in conjunction with inserted damp-proof courses, suggesting that such precautions are necessary to isolate decorations from previous accumulations of dampness in the wall, although it is obvious that water-proof plaster systems will also conceal any failure of the damp-proof course! It is usually suggested that tanking should not be used as it encourages condensation on wall surfaces, whereas normal permeable materials allow some diffusion of air to the exterior which reduces this condensation risk. In fact, the condensation risk is related mainly to temperature and, if the waterproofing is on the warm side of the wall structure, condensation is unlikely, and the risk of condensation is obviously reduced by ensuring that all materials below the waterproofing membrane are insu-lating, such as by the use of lightweight aggregate renders. Indeed, it is standard practice in timber frame buildings to provide an impermeable vapour barrier imme-diately beneath the internal plaster; this does not encourage interior condensation as it remains warm because of the excellent thermal insulation provided between this barrier and the exterior.

Dry lining is also sometimes used to remedy rising or penetrating dampness. Usually wood battens are fixed to the walls and covered with plasterboard, but simple dry linings are not entirely effective. Battens should be preserved and iso-lated by a damp-proof membrane from the background to avoid fungal decay, which is perhaps the most serious problem encountered, but it is not generally appreciated that dampness can accumulate behind dry lining due to condensation as warm humid air from the accommodation diffuses towards the cold wall supporting the battens. In theory, the best method for avoiding condensation risk is to ventilate the dry lining cavity to the exterior, the normal precaution against interstitial con-densation as explained in section 4.7, whereas often dry lining cavities are vented to the interior, actually encouraging interstitial condensation. If ventilation of the cavity to the exterior is considered to be unrealistic, the correct solution is to provide a vapour barrier such as polythene sheet immediately beneath the plaster-board to minimise diffusion of humid air from the accommodation towards the cold external walls, the normal precaution that is also used in timber frame walls.

Noise problems

5.1 Introduction

Sound problems in buildings may be broadly divided, for convenience, into acoustic and noise problems. Acoustic problems are those that arise in a unit of accommodation in connection with hearing, that is, the way in which the design and construction of the accommodation affects our ability to hear sounds which we wish to hear, such as speech and music. Noise problems are sounds originating elsewhere which interfere with our hearing or cause irritation. This chapter is concerned with these noise problems, considering acoustics only in relation to them.

When we are listening to a speaker or to music in an accommodation space, the sound passes directly to our ears. However, sound also passes to the room surfaces and is reflected, eventually reaching our ears from many directions. The distances travelled by the sound by these routes varies, so that we receive the sound from several directions at slightly different times, the reflections being called reverberant sound, and some of the routes involving reflections from several successive surfaces. Sound involves atmospheric pressure oscillations, the frequency of the oscillations giving the pitch of the sound and the amplitude of the oscillations giving the intensity or loudness. If sound from a source arrives by two different routes of different length, the oscillations will be out of phase; if they are half a wave out they will precisely cancel out to destroy the sound, but if they are one wave out they will reinforce and double the amplitude of the sound. This failure of coincidence of sound vibrations by different routes is known as interference and often causes serious distortion to speech and music. Interference can be used to judge the nature of a room, the degree of interference being expressed usually in terms of the reverberation time, which may vary from 0.5 seconds in a living room to more than 12 seconds in a large cathedral. If the time is less than about a second, reverberation will decay quickly with little interference with sound, but a long time will make it very difficult to understand speech, although it may account for the 'atmosphere'; in a cathedral we can only clearly hear very slow speech because of the long reverberation time, but the 'echoes' also account for the 'atmosphere' which is characteristic of cathedral music. Reverberation time depends directly on the volume of the accommodation unit, but indirectly on the sound absorption of the surfaces or the area of each different type of surface multiplied by its absorption coefficient; this represents the amount of sound energy that it can absorb at particular frequencies. Obviously, sound absorption is not limited to the structural surfaces but applies equally to introduced surfaces, so that the reverberation characteristics of an empty room are drastically magnified by furnishing, whether it is curtains and carpets or seats and tables, all of which increase absorption, reducing echoes and reducing reverberation time.

Basic information on acoustics in relation to speech is available in, for example, Building Research Establishment Digest 192, but the acoustics of music are much more complex. We accept to a certain extent that the musical atmosphere varies with the size of a room; the main design control required is usually in the nature of the reflecting surfaces, to ensure that the intensity of the reflected sound is not too intense, while sufficient reflection is retained to produce a reverberation time characteristic of the size of the building to maintain the characteristic atmosphere. Such problems are beyond the scope of this chapter and this book, and are the province of consulting scientists specialising in acoustics, rather than a consulting scientist specialising in building defects and deterioration. This chapter is therefore restricted to consideration of sound insulation requirements in buildings to limit external weather and traffic noise, or transmission of noise generated within normal domestic or office accommodation; noise generated, for example, by industrial machinery is not considered, although guidance is available in Building Research Establishment Digests 117 and 118.

There are two basic sources of external sound that affect persons in buildings. Sound generated in air as speech, music or other noise is absorbed into floors or walls which transmit the vibrations to air in other parts of the accommodation. In addition, direct impact on floors and walls will be transmitted to the air in other areas, although impact usually arises as footsteps so that insulation in this respect is only usually required for floors. Although direct transmission through floors and walls represents the main noise problem, flanking transmission can also occur in which sound in one area is transmitted through the structure to other more distant areas. For example, sound generated in a ground floor room in a terrace unit may be heard by direct transmission in the first floor room above or in an adjacent ground floor room in the adjoining unit, but the sound may also be heard in a first floor room in the adjoining unit through flanking transmission through the walls.

The transmission of sound involves transmission of energy, and the normal laws of physics indicate that the transmission efficiency will depend on the velocity squared, which depends on the amplitude squared. Doubling the amplitude or volume of the sound will therefore increase transmission by a factor of 4. However, the laws of physics also suggest that the transmitted energy will be inversely proportional to the square of the mass involved, so that double wall mass will reduce transmission by a factor of 4, a relationship that is often described as the *mass law*. The laws of physics also indicate that an increase in sound frequency will reduce sound transmission, so that it is easier to provide efficient insulation against high frequency sound.

Structural components, such as walls, also have a critical frequency at which they vibrate in sympathy with imposed sound, increasing sound transmission efficiency and therefore reducing sound insulation efficiency. The critical frequency depends basically on the mass per unit area for the structural element, the critical frequency being halved by doubling the mass, usually by doubling the thickness of the element. For example, typical brickwork with a thickness of about 210 mm has a density of about $400 \, kg/m^2$ with a critical frequency of about 100 Hz, whereas brickwork with a thickness of about 100 mm has a density of about $200 \, kg/m^2$ and a critical frequency of about 200 Hz. Similarly, gypsum plasterboard with a thickness of 10 mm has a density of about $9 \, kg/m^2$ and a critical frequency of 2900 to 4500 Hz, whilst plasterboard with a thickness of 20 mm has a density of about $18 \, kg/m^2$ and a critical frequency of about 1400 to 2300 Hz. If there is sufficient mass there will be sufficient insulation, even at the critical frequency, and this is normally the approach in traditional construction, but where lightweight construction is involved, whether it is the use of lightweight blockwork or timber frame construction, for example, it is necessary to design walls and floors to have critical frequencies either below or above the normal noise problem range of about 100 to 3200 Hz.

Whilst it might therefore seem attractive for lightweight construction to design for a critical frequency in excess of 3200 Hz, this is not a simple practicable solution as sufficient mass is still required to satisfy the *mass law* requirements for insulation.

The solution to this apparent contradiction is, of course, more complex structural design, particularly the introduction of cavities or double leaf construction. Efficiency of sound transmission is reduced, or sound insulation improved, with each interface where sound is transferred from air to a solid or solid to air. The insulation efficiency of a cavity is perhaps best illustrated by the fact that sound transmission between two rooms is obviously dramatically reduced if there is a third room between them! The sound insulation efficiency of a cavity increases with width, although there is little improvement beyond a cavity width of about 100 mm. This is a particularly important observation in relation to double glazing, as the use of a 100 mm gap can provide efficient sound insulation, despite the low structural density of double glazing. Similarly, for lightweight double leaf walls a cavity in excess of 100 mm would give optimum sound insulation; but this is usually unrealistic and smaller cavities must be used, usually 25–50 mm, although an absorbent material in these smaller cavities can greatly improve their effectiveness. Generally, glass fibre quilts and other low density thermal insulations are equally effective in this respect as sound insulations.

The sound insulation value of cavities and absorbent materials can be easily destroyed by bridging, cavity wall ties being particularly troublesome in this respect. In addition, whilst cavities and absorbents are effective insulators against relatively high frequency sound, such as high pitched voices and music, they do not necessarily give good insulation against low frequency sound, such as percussion music and impact sound such as footsteps, which can only be limited by the use of mass in the structure as previously indicated.

5.2 Noise within buildings

The Building Regulations for England and Wales have included sound insulation requirements for many years. The main problem is considered to be sound transmission between adjacent building units in flats, terraces and semi-detached dwellings, requiring the highest insulation that is considered to be practicable. It will be appreciated from the discussion in section 5.1 that the general requirement is for density in terms of mass per unit area of wall or floor, and where lower densities are required for structural or thermal insulation reasons, cavities must be introduced, perhaps filled with absorbent material. Thus, where solid walls are used, it is essential to obtain adequate mass by using dense materials or much greater thicknesses of lightweight materials. Where cavity walls are used it is important to avoid bridging with structural features or ties; floor slabs must not bridge cavities and in constructions where ties are necessary they should be stainless steel wire ties rather than heavier twisted strip steel ties. These comments apply to party walls between adjacent units, but there are no mandatory requirements for sound insulation of partitions within the units, although it is obviously desirable to provide a reasonable degree of privacy. A solid partition of brickwork or blockwork, plastered both sides and weighing 75–150 kg/m^2, is adequate in normal construction, but a double leaf stud partition of half this weight will also achieve adequate sound insulation if the cavity width is greater than about 50 mm.

Floor insulation seems to present greater difficulties, perhaps because floors are required to give resistance to both airborne sounds, such as conversation, and impact sounds, such as footsteps. Generally these requirements can be achieved in two different ways, either by providing sufficient mass or by using a floating floor, in which the floor is isolated from the structure by a resilient support; in practice a resilient floor covering such as a heavy carpet will be as effective. The stiffness of the resilient layer must be carefully selected; fibre material, such as glass fibre or mineral wool, is usually sold by density rather than stiffness so that an appropriate density must be selected. Increasing the thickness of the resilient layer will improve acoustic insulation, particularly in relation to low frequencies, but if the layer is too thick a floor may feel soft and, with a floating screed construction, the screed may be susceptible to cracking. A typical fibre resilient layer for a floating screed would have an uncompressed thickness of 13–25 mm and density of at least 36kg/m^3; expanded polystyrene is also suitable provided that an appropriate pre-compressed grade is used. Screeds must not be less than 65 mm thick to minimise the danger of cracking; floor screeds are discussed in more detail in section 14.2.

Acoustic insulation requirements can also be achieved by constructing wood flooring on a concrete slab base, provided that the floor battens are isolated from the concrete base using a suitable resilient support, such as glass fibre or mineral wool quilt, or by supporting the battens on proprietary acoustic clips with quilt laid between the battens to minimise any airborne sound transmission between the floor and the concrete slab base.

Changes in specification can lead to unexpected problems in performance. In one case a chipboard floor laid on battens was required to be isolated from the support-ing concrete base using proprietary acoustic supports with glass fibre quilt laid between the battens. However, the specification was varied during construction with the battens laid direct over a continuous quilt; problems were encountered with levels, foot traffic over the floors produced considerable noise at the tongue-and-groove board joints and the floor support felt spongy in places. The contractors attempted to remedy the situation by using wedges of various materials to level the battens, but these were unstable and the contractors then provided angle brackets fixed between the battens and the concrete floor slab. These brackets generated further noise, as they were standing on top of the glass fibre quilt but fixed with shot nails through the quilt into the concrete slab beneath, so that foot traffic caused the brackets to move on the shot nails and generate additional noise. Whilst the wedges and brackets achieved reasonable levelling of the floors, the local loads on the wedges and the brackets also provided very efficient acoustic bridges between the floor and the concrete slab beneath, completely destroying the acoustic insula-tion properties of the resilient glass fibre quilt. Whilst complaints were concerned with the noise generated when walking over the floors, it soon became apparent that this noise was being transmitted to other flats in the same complex, particularly through concrete floor slabs to the flats beneath, but also by conductivity through the floor slabs and reinforced concrete frames of the building to adjacent flats.

Sound insulation can be provided in wood joist floors, provided that sufficient attention is given to the necessity to provide adequate weight coupled with optimum utilisation of the floor cavities. For example, two suitable floor constructions are described in Building Research Establishment Digest 266. In the first form of con-struction a resilient layer such as glass fibre quilt is laid over the joists and supports a raft comprising battens, 19 mm plasterboard and finally 15–22 mm chipboard. The ceiling beneath comprises two layers of plasterboard, 12.5 mm and 19 mm thick. It is important in this form of construction to ensure that the floor raft is totally iso-lated from the structure by the resilient layer, so that the battens must not be nailed to the joists and the resilient layer must be turned up at the edges of the floor to isolate the edges of the floor raft from the adjacent walls. In the second form of con-struction, usually known as a platform floor, a platform is laid on the joists followed by a resilient layer, and finally a floating floor surface, usually of chipboard; again the resilient layer must be turned up at the edges to prevent the floating layer from contacting the walls. In a raft type construction, the load is concentrated on the resilient layer passing between the battens and the supporting joists, and the resilient layer must therefore have adequate density to provide support without excessive compaction; mineral wool 25 mm thick with a density of about 140 kg/m^3 is suitable. With a platform construction the load is spread over the whole area of the resilient layer which can be less dense; 25 mm thick mineral wool with a density of about 70 kg/m^3 is suitable. In most cases the performance is improved by adding 100 mm of absorbent material, such as glass fibre quilt designed for roof insulation, between the joists.

5.3 Wind noise

Wind noise can be troublesome, particularly in severely exposed situations. Wind noise typically has two components: a high frequency whistle which must be insulated from the accommodation using the same techniques that are employed for preventing speech and music noise from being transmitted between adjacent living units, and low frequency buffeting which must be insulated in the same way as for impact sound, such as footsteps in adjacent living units. High frequency wind noise or whining is caused by air passing through or over orifices and generating characteristic oscillations in the same way as in familiar musical wind instruments. Whilst careful draught exclusion can limit noise from air penetration through windows and doors, little can be done about noise generated by air flow over or through a roof, and the most satisfactory solution to the problem is the use of absorbent insulation in the roof space, usually on the ceiling. As glass fibre quilt thermal insulation performs well as a high frequency sound absorbent, the necessary insulation is usually provided in any case in modern buildings.

Low frequency noise is caused by the disturbed air flow round and over the building which causes low frequency vibrations within the structure. Wind buffeting noise is inevitable within conventionally constructed buildings in exposed positions as there are many noise conduction routes between the external envelope and the inner accommodation structure. Obviously solid walls transmit sound better than cavity walls, but rigid heavy steel cavity ties transmit the noise much more efficiently than wire butterfly ties. Most of the low frequency noise is generated around the roof, particularly at the eaves. In theory, noise can be avoided in the accommodation by isolating the roof structure, as in the floating floors described in section 5.2, but in practice it is impossible to provide a suitably resilient mounting for a roof which is also capable of resisting wind lift forces. The best solution is therefore to design a roof to minimise the turbulent air flow that is responsible for the noise. Moderate pitch roofs are more efficient than either flat or steeply pitched roofs. Hipped roofs are more efficient than gable roofs, and flush eaves are more efficient than projecting eaves. However, these preferences must be balanced against stability and weather resistance requirements. In practical terms, the best solution is to choose a form of roof that will be appropriate to the exposed weather conditions in terms of projecting eaves and slope, and then to minimise the noise generation by using a hipped rather than gable structure, as well as by using smooth interlocking tiles rather than pantiles and other irregular shapes which interfere with air flow and cause turbulence.

5.4 Traffic noise

Air traffic noise usually causes more irritation than road or railway noise, even if the noise is less intense. This exaggerated reaction to aircraft noise is probably due, at least in part, to the general public attitude that air traffic is a new development and avoidable, a reaction sometimes described as the NIMBY (not in my backyard) syndrome, whereas road and rail traffic is generally established and cannot be avoided in this way; in fact, reactions to new motorways and the new Channel Tunnel rail link are exactly similar to the usual reactions to air traffic as they are also considered to be avoidable problems! These are obviously psychological and political problems, but there are other differences between air and surface traffic.

Aircraft noise affects a very wide area and persists for a significant time, despite the high speed of modern aircraft. In addition, both propjet and particularly jet aircraft produce high frequency sound which many persons find particularly irritating; it is often described as 'penetrating' noise. In fact, aircraft also produce low frequency noise, particularly when climbing away from an airport at relatively low speed when jet propelled aircraft in particular suffer from severe turbulent flow around the aircraft exhausts. In contrast, surface transport produces mainly low frequency noise or 'rumble' which tends to cause vibration of the structure rather than distinctive noise in many instances, high frequency noise being restricted to very high speed motorways and express rail tracks. As all types of traffic noise involve both high and low frequency sound the sound insulation principles described in section 5.1 apply, although the type of insulation is usually biased in relation to the type of noise that is considered to be most troublesome.

The general method of protection against traffic noise is, of course, the establishment of an enclosure. The mass of an envelope is very important in terms of general insulation from external sound, and lower density areas such as doors, windows and roofs represent zones through which sound can penetrate more readily. It is therefore necessary in traditional brickwork or blockwork buildings to concentrate particularly on the susceptible zones which are orientated towards the sound source; in lightweight structures the same principles apply, except that it is necessary to consider the entire elevations facing the sound source.

It will be appreciated from section 5.1 that mass is very important, but absorbing materials can also be used, particularly to minimise high frequency noise. Where reasonably efficient insulation against air traffic noise is required, double glazing is much more efficient than single glazing, but the thickness and mass of the glass are almost as important as the absorbency related to the width of the air gap. Reasonably efficient insulation, usually acceptable for grant purposes, can be provided using double glazing comprising:

3 mm (24 oz) glass, minimum 200 mm (8") air gap, or
4 mm (32 oz) glass, minimum 150 mm (6") air gap, or
6 mm ($\frac{1}{4}$") glass, minimum 100 mm (4") air gap.

External doors must also be insulated, either by the use of similar double glazing or a hollow panel door filled with absorbent material, although a heavy wood door will be just as effective. Window and door insulation alone is often adequate for surface traffic noise, although it will be appreciated that ventilation may present a problem as opening the windows will completely destroy the sound insulation effect; even ventilators with limited area will permit noise penetration. It is therefore better to provide sound barriers in the form of specially designed fences or earth banks to isolate areas of housing from troublesome motorways and railways, a problem that is considered in more detail in Building Research Establishment Digests 185 and 186. However, in the case of aircraft noise, or even noise from roads or railways on high embankments, lightweight roof structures may also require insulation. Whilst some improvement in the insulation can be achieved by introducing absorbent material on the ceilings, an improvement that is automatically achieved with adequate and efficient thermal insulation, other improvements are only possible through the selection of appropriate roof types in new construction. A flat concrete roof with a density of 200 kg/m^2 or more, usually provides adequate insulation because of its mass. A timber roof is much less effective, although the insulation value is much greater if there is a separate ceiling beneath and it can be improved further with absorbent material on top of the ceiling. A pitched roof is more efficient than a flat roof because of the larger airspace involved, and a pitched roof with adequate insulation on the ceiling can be almost as efficient as a flat concrete roof. However, roof lights present particular problems; if they light general areas, such as corridors or staircases, a higher level of noise may be tolerated, but if they light an accommodation area they must conform to the normal requirements for windows.

In general it must be appreciated that sound 'leaks' through any gap in the structural envelope, including flues, vents and open windows, but to a lesser extent through low density parts of the structure such as closed windows and roofs.

6

Wood problems

6.1 Introduction

Structural overstressing of wood components in buildings, perhaps causing distortion or collapse, is usually due to failure to observe established design and construction criteria, currently represented by British Standard Specifications and Codes of Practice and, for example, the Approved Documents which are published in support of the Building Regulations for England and Wales. Inadequate dimensions in relation to structural requirements, that is in relation to loads, spans and component spacings, can be readily checked and need no special comment in this book, other than a reminder of the basic principles involved; in relation to cross section, the strength of a component depends on the width and the square of the depth, and when designing to support a particular load the required strength is proportional to the square of the span, remembering that the load is not the load density that is often quoted as a requirement but the product of the load density times the span. However, it is much more difficult to check that species and grading requirements have been observed, and this subject is therefore considered in more detail in section 6.2.

Moisture is profoundly important in relation to wood components, causing loss of strength, dimensional changes and distortion, as considered in more detail in section 6.3, but moisture content is also a major factor in relation to susceptibility to biodeterioration, as considered in more detail in sections 6.4 and 6.5.

Chemical deterioration of wood is much less common than biodeterioration and, as a result, less readily recognised, as discussed in section 6.6. Similarly, wood preservation is adopted as a means to avoid biodeterioration, but the problems caused by wood preservation itself are less readily understood and therefore discussed in section 6.7.

6.2 Wood grading

Structural design criteria for wood components assume that the wood has adequate strength, but this will depend on the species and grading of the wood used. In the system used in British Standard BS 5268:Part 2:1984 *Code of Practice for permissible stress design, materials and workmanship*, design flexibility is allowed by relating design criteria to strength classes, with the strength classes separately related to wood species and grading. Thus the widely used European redwood and whitewood can meet strength class SC3 with grades GS or M50; in British Standard BS 4978:1973 *Specification for timber grades for structural use*, GS represents General Structural grade determined by visual stress grading and M50 is the equivalent grade determined by machine stress grading. However, for weaker woods, more stringent grading requirements are necessary to meet the same strength class SC3, so that western whitewoods from the USA must be SS grade, and whitewood grown in the UK as European spruce and sitka spruce must be M75 grade.

Stress grading is not concerned with the characteristic strength of a particular species, but is concerned with selection for quality in relation to features that affect strength. The first feature in grading is to ensure that the pieces of wood possess their proper cross-section dimensions and any loss of dimension, perhaps through re-sawing, will result in a loss of strength; the permitted tolerances are defined in a separate British Standard BS 4471:1987 *Specification for sizes of sawn and processed softwood*. The presence of wane, that is the curved edge of the log, along the arris of a piece of wood also represents a loss of cross-section and thus a loss of strength; the area lost by wane from an individual face of a piece must be less than one quarter of the total area of the face for SS grade and one third for GS grade; in GS grade up to half the width can be lost in wane within 300 mm of the end of the piece, provided that the continuous length of the wane is less than 300 mm. However, the most important factor in grading is the Knot Area Ratio, defined as the ratio of the sum of the knot area projected on a cross-section relative to the total area of the section. The point of this requirement is that it is considered that the integrity of the cross-section throughout the length of the piece is perhaps the most important factor in terms of strength. However, when a piece is used as a beam, the upper wood is in compression and the lower wood is in tension. The margins of a piece of wood are defined as the outer quarters of the width of the piece and, when knots occur in these margins, they clearly have a greater significance on the strength of a beam than when they occur in the relatively unstressed centre of the beam. A margin condition is said to exist when more than half the area of either margin is occupied by projected knots. GS grade is then defined as wood with a Knot Area Ratio of less than one half or, if margin condition exists, less than one third; margin condition is not applied to square section pieces for which the Knot Area Ratio should be less than one third. SS grade has a Knot Area Ratio of less than one third or, in the case of margin conditions or square sections, less than one fifth. In estimating the Knot Area Ratio for grading purposes, all knots of less than 5 mm diameter are ignored and there is no distinction between empty knot holes, dead knots or live knots. The longitudinal separation of the knots is important, and for both GS and SS grades the piece must be rejected when two or more knots or groups of knots with a Knot Area Ratio of more than 90% of the total permitted are separated in length by less than half the width of the piece.

This grading system is obviously rather complex but it can be operated reasonably reliably if the grader adopts a logical sequence of decisions:

Knot Area Ratio (BS 4978)

less than $\frac{1}{5}$	SS grade
less than $\frac{1}{3}$	
no margin condition	SS grade
margin condition	GS grade
less than $\frac{1}{2}$	
no margin condition	GS grade
margin condition	reject
more than $\frac{1}{2}$	reject

Splits, fissures and resin pockets must also be taken into account. Unlimited fissures are permitted if they are less than one half the thickness of the piece. If they are greater than half the thickness but less than the whole thickness, their length should not exceed 900 mm for GS grade or 600 mm for SS grade, or one quarter of the length of the piece, whichever is the lesser. If the size of the defect is equal to the thickness, the length shall not exceed 600 mm for GS grade or, if the defect occurs at the end of the piece, its length should not exceed 1.5 times the width of the piece; in SS grade, fissures equal to the thickness are only permitted within one width length of the end of the piece.

The growth rate should not be excessive and there should not be less than four rings per 25 mm. In addition, the slope of the grain should be less than 1 in 6 for GS and 1 in 10 for SS grades; the angle of grain is determined by pulling a needle on the end of a shaped rod across the surface of the wood, although a method which is simpler but not mentioned in BS 4978 is to place a spot of Indian Ink on the surface of the wood as it will spread along the grain, indicating its slope.

Finally, the wood should not be excessively distorted. Within a length of 3 m, bow should not exceed one half the thickness, spring should not exceed 15 mm and twist should not exceed 1 mm per 25 mm of width. In addition, the depth of cupping should not exceed $\frac{1}{25}$ of the width. Some insect borer damage is permitted, such as pinholes, provided there is no evidence of current activity, but larger holes, such as those produced by wood wasps, result in rejection. Fungal decay damage and other abnormalities also result in rejection; sap stain is not structurally significant and is not considered in stress grading.

In machine stress grading all these visual factors are ignored, except for decay and excessive distortion, as the wood is passed between rollers which deflect it and automatically measure its elasticity in bending as a means of grading. Obviously visual stress grading according to these rather complex British Standard rules is only economic when graders are very experienced, but grading machines have the advantage that they only require conscientious operation rather than extensive grading experience.

The Building Regulations for England and Wales actually permit the use of wood which has been graded to other systems such as the Canadian NLGA rules, but only in the sense that grades are selected which match the British Standard grades; if it is necessary to grade wood that is already in service in a building it is only necessary to consider the British Standard grading rules and that is why they have been quoted in detail in this section.

If a wood component does not meet the strength class requirements in the design, either because a weaker species has been used or a lower grading, there is obviously a risk of excessive distortion or collapse. Species identification is not easy with the softwoods that are generally used for structural purposes in the British Isles, but grading is not too difficult once the basic principles of determining Knot Area Ratio have been understood. A particularly important point is margin condition where knots occur in the outer top or bottom of a piece in service. It will be appreciated that, if wood is re-sawn to a smaller size, central knots which were originally of limited significance may become more important if they then represent margin condition; many of the doubts on grading in new buildings arise through resawing in this way, and the failure to appreciate that resawn wood must be regraded in the same way as the original sawn wood. Imported wood is usually reliably and consistently graded, and problems are usually associated with home grown wood as producers are not yet generally experienced in grading techniques, or even perhaps aware of the need for grading; it is only in recent years that home grown woodlands have been able to produce significant quantities of timber of structural dimensions.

Grading is equally important in assessing the strength of timbers in old buildings as in new buildings, as the strength requirements are the same in both cases, based as they are on structural safety. Nineteenth century softwood buildings do not present too much difficulty but medieval buildings, for example, are more difficult to assess, partly because of the unusual cross section dimensions of the timbers and the design of the supporting joints, but also because oak and chestnut are used rather than softwoods. The fact that a building has survived is sufficient evidence that it is structurally sound, assuming that it has not suffered local insect borer or fungal deterioration which affects its strength, but difficulties arise when old buildings are being converted for modern use which may involve excessive floor loadings. In such instances it is necessary for engineers to calculate strength rather than to rely on standard tables, while still using Strength Class principles and thus the normal grading procedures in BS 4978 which have been summarised previously in this section.

6.3 Moisture content

The properties of wood are profoundly influenced by the presence of water. Freshly felled wood is often described as green and has a very high moisture content varying typically from 60 to 200%. The green moisture content tends to vary inversely with the normal dry density of the wood, so that black ironwood with a specific gravity of 1.08 has a green moisture content of only 26%, whereas South American balsa with a specific gravity of 0.2 has a green moisture content of about 400%. During drying free moisture is first lost from the cell spaces and this involves little change in properties except for a change in density. Eventually drying will result in the wood reaching the fibre saturation point, normally at a moisture content of 25–30%, and further drying results in the loss of bound moisture from the cell walls.

The loss of bound water typically results in shrinkage and a progressive change in physical properties. In addition, the amount of bound water remaining in the wood is approximately proportional to the relative humidity of the surrounding atmosphere; wood is hygroscopic and will gain or lose water with changes in atmospheric humidity, and these changes will result in swelling or shrinkage. In fact, the moisture content of the wood tends to lag behind the changes in relative humidity of the atmosphere, a phenomenon known as hysteresis illustrated in Figure 6.1. Most of these changes in properties with variation in moisture content can be attributed to the sub-microscopic or chemical structure of wood.

Swelling or shrinkage with changes in moisture content is known as movement. In the longitudinal direction the movement is generally very low, only about 0.1% for a change of moisture content from normal dry wood at about 12% to the wet

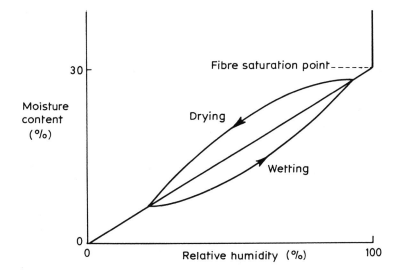

Fig. 6.1 Variation in the moisture content of wood with changes in atmospheric relative humidity.

condition at fibre saturation point or above. In contrast, the movement in the radial direction for the same moisture content change is about 3–5%, and in the tangential direction 5–15%. The difference in movement between the radial and tangential direction can be largely attributed to the fact that early or spring wood shrinks less, so that in the radial direction the movement can be attributed to the average of the early and late wood shrinkage, whereas in the tangential direction the dense and strong late wood tends to control the physical behaviour and thus generate higher movement. Indeed, if a piece of softwood is planed smooth at a low moisture content and then wetted, the late wood will swell to a greater extent than the early wood to give a corrugated surface, usually known as 'washboarding'.

Generally a higher lignin content in a particular wood species will result in lower movement or greater stability. The high shrinkage in the tangential direction, coupled in some species with weakness induced by large medullary rays, sometimes results in surface splits developing during drying. Movement characteristics can be largely attributed to the microfibril angle in the cells, and as this angle changes progressively from the pith outwards there is also a progressive change in movement. Heartwood close to the pith is most stable but, in addition to this progressive change, there is a far larger alteration in movement between heartwood and sapwood in many species, the heartwood being restrained by the extractives with which it is naturally impregnated and sapwood often possessing extremely high movement. These changes in movement relative to the original position in the tree, coupled with distorted annual rings and twisted grain, can result in warping or splitting, defects that will be described in more detail later in this section.

Water vapour is lost or gained most rapidly through end-grain surfaces because of their high permeability, a factor which also encourages very rapid absorption of liquid water. This means that many defects arising through moisture content changes, such as splits and the development of decay or stain fungi, are concentrated around end-grain surfaces. In general, wood is used as long pieces with only small end-grain surfaces so that the side-grain is probably more significant, particularly in service in dry conditions where it is the hygroscopicity of the wood, coupled with changes in atmospheric relative humidity, that are likely to have the greatest effect. Wood softens as moisture content increases towards the fibre saturation point, a property that is very valuable when bending wood or peeling veneers from logs; however, it substantially affects strength, so that in structural uses where wetting may occur wood must be employed in adequate dimensions to tolerate the strength loss that will occur as the moisture content approaches the fibre saturation point.

When trees are first converted to sawnwood the high moisture content introduces a danger of fungal damage, quite apart from the excessive transport costs that are involved if the weight of the wood is perhaps doubled by an excessive moisture content. However, if green wood is used in buildings the most serious problem, other than the danger of fungal decay, is the cross-grain shrinkage that will occur as the wood moisture content gradually achieves equilibrium with the relative humidity of the surrounding atmosphere. This is a very serious problem in certain uses, such as window and door frames, floor boarding, panelling and other uses where cross-grain shrinkage will result in unacceptable gaps between individual pieces of wood. It is therefore usual to air-season or kiln-dry wood before installation to the average moisture content that it will encounter in service. The main problem is that wood may be processed to the required moisture content, but this may change during storage, delivery or installation, particularly in buildings where the wet trades, such as brick laying and plastering, cause very high humidities during construction. Indeed, floors are sometimes laid in new buildings very promptly after kiln-drying; the high humidity then causes them to expand so that the entire floor may lift in a dome or the expansion causes serious damage to the surrounding walls.

Stacking the wood for a period inside the building before laying will avoid this particular problem, as it will reach equilibrium with the high relative humidity in the damp new building but, when the building is ultimately completed, the humidity will fall and the wood will shrink. The integrity of the floor, and thus its ability to act as a fire break or thermal insulator, is naturally affected by this shrinkage, unless the boards are joined by tongues to span the shrinkage gaps.

A floor is, of course, a rather large panel, normally assembled from separate boards, and the problems encountered in floors are naturally encountered in other panels, such as wall linings, doors and furniture. Several panel materials are now available in which wood has been processed to minimise these shrinkage defects. In plywood the wood has been cut into veneers which are orientated at right angles to one another so that the stable longitudinal grain is used to restrain physically the movement in the cross-grain direction. This method for reducing shrinkage is particularly successful in normal buildings but sometimes fails under extreme conditions, such as when used in boat building. Adhesives have been developed which will withstand the very high stress that is developed when the moisture content fluctuates from fibre saturation point down to low levels in very dry conditions; if this occurs regularly, however, there is a tendency for the wood to rupture on either side of the glue line, and this can be avoided only by using wood of moderate or low movement in the manufacture of the plywood. It should be appreciated that the plywood still expands normally in thickness, but this is seldom a problem unless it affects the weather seal when plywood is used as exterior panelling.

One particular advantage of plywood is that, as the grain runs horizontally in both directions within the plane of the board, it is equally strong in both directions. In contrast, it could be said that the alternative process for producing a panel, the manufacture of particle board, results in a material that is equally weak in both directions. In particle board manufacture, the principle is to divide the wood into small particles which are then randomly orientated so that the movement is shared equally in all directions, although it is restrained to some extent if sufficient adhesive is used. The movement is higher than for plywood and, whereas plywood presents a completely smooth surface, a particle board is comparatively irregular, a defect that is particularly noticeable with changes in relative humidity as the surface chips expand and contract across the grain. Smaller particles are sometimes used for the surface of the board, or the surfaces are veneered in attempts to minimise this problem. Unfortunately veneers are often applied on boards with comparatively large surface chips in which any movement results in deflection of the veneer or 'shadowing'.

The only alternative is to reduce the wood to individual fibres which are then randomly orientated and reconstructed into a board which is, in effect, a rather thick sheet of paper. Low density boards are known as insulation boards and are used for lining walls and ceilings, but the medium and high density boards (MDF and hardboard) present a reasonably hard and smooth surface for painting. Whilst fibre boards might appear to be similar to plywood in principle, although the wood components are re-orientated on a rather different scale, they do not achieve the same end results. In fibre boards the movement tends to be randomised in all directions, whereas in plywood the movement in the plane of the board is physically restrained. Plywood is therefore preferred for external panelling and other situations where it may achieve a high moisture content, as its movement within the panel dimensions remains small, whereas the significant swelling of fibre board in similar conditions may lead to serious distortion.

Fibre saturation point usually occurs at a moisture content of about 25–30%; movement occurs in wood only at lower moisture contents. Air-drying will normally reduce the moisture content to 17–23%, but if lower moisture contents are required kiln-drying is essential. Wood can be pressure-treated with preservatives at moisture contents below 25%, but for use in buildings wood should be dried to the average moisture content that it will achieve in service. Carcassing or framing timber in buildings can tolerate a relatively high moisture content, largely because the cross-section dimensions are relatively unimportant, but drying is necessary to prevent fungal decay; fungal infections will only develop on wood at moisture contents above about 23%, but Dry rot infection can spread from adjacent infected wood at moisture contents as low as 18% in poorly ventilated situations, such as the rear surfaces of skirting boards, and it is therefore advisable that all wood in buildings should be dried to below 17%. In fact, lower moisture contents are usually necessary to ensure freedom from shrinkage defects. In a house with central heating, the average wood moisture content will be about 12%, although perhaps 14% in bedrooms with their lower heating levels. In buildings which are more intensively heated, such as offices and hospitals, the typical wood moisture content may be as low as 8%, and perhaps even less in flooring installed on floor heating. In climates with very cold and dry winters, the atmospheric relative humidity can be very low and results in moisture contents as low as 4% in buildings with central heating, yet in the spring the high humidity may increase the moisture content to as much as 12%, so that the choice of a preferred moisture content for kiln drying is rather difficult. Indeed, these required moisture contents are all largely theoretical; it is very difficult to ensure that wood remains completely protected during the numerous storage, transportation and handling stages between leaving the kiln and ultimate use in a building. The worst stage is perhaps installation in a new damp building, in which any wood will automatically be reconditioned to a higher moisture content appropriate to the surrounding conditions.

Warping is the general term for distortion on drying and takes several forms. Cupping is warping across the grain or width of a board and arises in flat-sawn boards through the outer zones of the trunk, particularly the sap wood, shrinking to a greater extent than the inner zones, giving a planoconcave surface on the outer side of the board relative to the trunk. Flat-sawn boards used for flooring should always be laid with the heart upwards, to give a planoconvex surface on drying and thus avoid trip edges. The alternative is to use more expensive quarter-sawn boards which are resistant to cupping and also more hard wearing. Twisting arises in boards through the presence of spiral grain and is really the result of wood selection rather than a drying defect, although spiral grain is quite common and perhaps unrealistic to reject. Bowing is longitudinal curvature arising perpendicular to the surface of a board, either through spring in a flat-sawn board or, more usually, as a result of sagging in the stack during drying through inadequate support or 'sticking'. Spring or crooking is longitudinal curvature within the plane of the board caused by tensions within the original tree which are released by conversion sawing, and can be severe in some species, giving decreased strength. Finally, diamonding occurs in pieces of wood of rectangular section when the annual rings pass diagonally across the end-grain; the radial shrinkage on drying is less than the tangential shrinkage so that the diagonals in, for example, a square section piece of wood become different in length and give a diamond shape.

Although it is normal to dry wood to a moisture content equivalent to the average atmospheric relative humidity anticipated in use, it is common to encounter movement problems. Faults, such as gaps appearing between floor blocks or boards, are due to the wood drying after installation, either through inadequate kilning or usually through re-wetting between kilning and installation. A door or drawer jamming in humid weather may be slack in drier conditions. Frames which introduce an end-grain surface in contact with side-grain will inevitably result in cracking of any surface coating system. In other situations the cross-section movement may become apparent as warping. The obvious solution to all these problems is to use only wood with low movement, but this is not always realistic. The only alternative is to impregnate the wood with chemicals which induce stabilisation, although processes of this type are also frequently unrealistic because of their cost and the difficulty in achieving deep penetration.

It may seem that the only realistic solution is to enclose wood within a protective film to stabilise the moisture content. Paint and varnish coatings will act in this way, provided they completely cover the wood and they are not damaged in any way. However, whilst these coatings will give good protection against rainfall, they are unable to prevent moisture content changes resulting from slow seasonal fluctuations in the surrounding atmosphere, and it is the slow movement changes that result in this way that cause surface coatings to fracture wherever a joint involves stable side-grain in contact with unstable end-grain; this is the most serious problem with wood window frames. Rain is absorbed by capillarity into the fracture in the paint, yet the remaining paint coating restricts evaporation, so that the moisture content steadily increases until fungal decay inevitably occurs if non-durable wood is involved. It is usually suggested that preservation provides a simple solution to this problem by reliably preventing decay, and most modern wood window frames are preserved, usually by double vacuum treatment; but preservation will only protect the wood and it will not stop water from damaging the paint coating. Wood is hygroscopic and covered with hydroxyl groups which have a strong affinity for water, so that water accumulating beneath a coating by condensation or penetration through paint fractures will tend to coat the wood elements, displacing paint and varnish coatings. This failure occurs because the wood prefers to be coated by water rather than by oil, so that this form of failure is usually known as 'preferential wetting'; it is responsible for blistering and peeling in paintwork and loss of transparency in varnishes.

The rate and extent of preferential wetting failure can be reduced by designing paint and varnish coatings to avoid interstitial condensation within wood. An impermeable paint coating can be used on the inner surfaces of the frames to reduce diffusion of warm humid air from the accommodation into the wood, with a permeable coating at the exterior which will prevent rain absorption but will allow any accumulating moisture to disperse by evaporation. Unfortunately, some limited preferential wetting failure still occurs and coating failure can only be avoided by introducing a permanent bonding system between the paint coating and the wood substance. The traditional approach is to use a primer coat which will penetrate deeply and establish mechanical bonding, but this system delays rather than prevents preferential wetting failure; this means that it is still necessary from time to time to completely strip paint from external wood surfaces, this work being the most expensive part of regular painting maintenance. The only alternative is to introduce a primer system that will chemically bond the paint to the wood, so that maintenance then consists simply of touching up physical damage and cleaning of the surface with sandpaper or solvent before applying a maintenance coating to preserve the appearance of the finish. Lead primers were advantageous because they achieved a degree of chemical bonding in this way through the formation of lead soaps, but these systems are no longer permitted. About 20 years ago the author patented a process to achieve permanent chemical bonding by the use of organic compounds of Group IV metals such as lead, tin and silicon, but the process was not well received by the coating industry, apparently because they were not keen to lose sales of primers and undercoats for maintenance painting which apparently represented a very large proportion of their business! Paint and varnish coatings for wood are discussed in more detail in section 9.2.

6.4 Fungal infections

If wood becomes damp it will support a wide variety of fungal infections. Many of these infections are superficial and do not affect the strength or integrity of the wood, such as the surface moulds which form green or black, occasionally yellow, powdery growths which are easily brushed or planed away. In fresh green wood the residual sap encourages the development of sapstain fungi, also known as bluestain, as the dark-coloured hyphae running through the wood produce most commonly a greyish-blue coloration. Sapstain fungi do not usually affect the structural strength of wood; however, by attacking the cell contents, staining fungi and invisible bacteria developing in similar conditions can make the wood more porous, allowing it to absorb water more readily and perhaps encouraging the development of other fungal infections in some conditions. If wood remains very wet continuously, soft rot fungi are likely to develop causing softening of the surface of the wood to a progressively increasing depth. However, if intermediate dampness exists, in which the wood has an adequate moisture content without being water-logged so that there is free access to oxygen in the air, wood destroying fungi can develop.

Generally, staining and soft rotting fungi are encountered in buildings only when they have been introduced as infected wood, although there are some unusual situations where soft rot may occur, such as in flooring affected by continuous severe dampness. The superficial moulds do not confine themselves to wood but occur also on damp plaster, wallpaper and carpets; their main significance is that their presence indicates very damp conditions. The wood-destroying fungi occurring at intermediate moisture contents, mainly Basidiomycetes, are the only group of fungi that cause severe deterioration of wood in buildings. Most of these fungi are Wet rots able to develop only when wood is actually wetted by moisture from another source, such as a leaking roof or plumbing, or by absorption of moisture from the soil or from other damp structural materials. With Wet rots the removal of the source of moisture is sufficient to prevent further deterioration, but preservation treatment may be necessary as a precaution against development of infection through similar wetting in the future, or because the drying stages following Wet rot infection may encourage the development of the Dry rot fungus.

The Dry rot fungus *Serpula lacrymans*, formerly known as *Merulius lacrymans*, presents special problems for several reasons. Once this fungus has become established it is able to generate moisture through digestion of wood so that it can maintain the atmospheric relative humidity under poorly ventilated conditions, permitting infection to spread actively even in the absence of a source of moisture. Dry rot infections are therefore often confined to poorly ventilated areas, such as behind skirtings or beneath floors, and the damage becomes apparent only when the visible surfaces become distorted or collapse occurs. In addition, this fungus is able to develop rhizomorphs or conducting strands which can convey water and nourishment, enabling the infection to spread across and through inhospitable materials, such as plaster and masonry, in the search for wood. It is therefore important to distinguish between the Dry rot fungus and the various Wet rot species which cannot control conditions or spread in this way. The following general descriptions of wood-destroying fungi include species most commonly found in the British Isles and will be sufficient for most purposes; more detailed descriptions are given in the books *Wood Preservation* and *Remedial Treatment of Buildings* by the same author, the latter book including a diagnostic table at Appendix 2.

Whilst the Basidiomycete fungi have already been divided into Dry and Wet rots, it is also convenient to divide them into Brown and White rots, depending on the manner in which they destroy wood. In a Brown rot the fungal enzymes destroy the cellulose, but leave the lignin largely unaltered, so that the wood acquires a distinct brown colour. As decay progresses the wood appears very dry and shrinkage cracks appear both across and along the grain, the size and shape of the resulting rectangles being a useful feature in identification. In contrast, the White rots destroy both cellulose and lignin, leaving the colour of the wood largely unaltered but producing a soft felty or stringy texture.

The Dry rot fungus *Serpula lacrymans* is certainly the best known wood-destroying fungus. The conditions for spore germination are extremely critical, involving prolonged exposure to an atmospheric relative humidity of about 95% on wood with a moisture content of about 25%. Spore germination is encouraged by acid conditions. This combination of circumstances usually occurs when a leak has caused wetting and Wet rot development, but repairs then result in progressive drying which is very slow in wood with restricted ventilation, such as the backs of painted skirtings and built-in lintels and beam ends, the earlier Wet rot damage causing the acidity that encourages subsequent Dry rot development. Once established the fungus will digest wood and produce sufficient moisture to maintain development, even in the absence of a source of moisture, provided that ventilation is restricted. Dry rot can also develop where a source of continuing dampness causes a moisture content gradient through wood components, with Wet rots infecting the areas of high moisture content and Dry rot infecting areas of lower moisture content; this form of fungal decay development is usually associated with a continuous plumbing leak, commonly caused by a defective joint or a floorboard nail driven through a central heating pipe.

The Dry rot fungus produces white hyphae which are, in fact, very fine tubes or hollow threads which progressively branch and increase in length so that they spread in all directions from the initial point of germination, provided that a food source is available. As food is exhausted some hyphae are absorbed, whilst others are developed into much larger rhizomorphs or conducting strands able to transport food and water. Exploring hyphae finding no nourishment are absorbed to form food for growth in more promising directions, giving the fungus the appearance of sensing the direction in which to spread towards a food source. Seasonal changes sometimes inhibit growth which then resumes when suitable conditions return. In this way hyphae compact on drying to form thin sheets or mycelium, each successive layer of mycelium indicating a season of growth. Active growth is indicated by masses of hyphae like cotton wool, perhaps covered with 'tears' or water drops in unventilated conditions, which enable the fungus to regulate atmospheric relative humidity and which also explain the Latin name *lacrymans* ('weeping'). Rhizomorphs may be up to 6 mm in diameter and can extend for considerable distances over and through masonry and behind plaster, spreading through walls between adjacent buildings and providing a residue of infection, even if all decayed wood is removed. Treatment of adjacent sound wood and replacement of decayed wood with preserved wood must always be accompanied by sterilisation treatment of infected masonry.

The mycelium is initially white, but then greyish and later yellowish with lilac tinges if exposed to light, and often subsequently green in colour through the development of mould. Sporophores or fruiting bodies generally develop when the fungus is under stress through shortage of food or moisture. Sporophores are shaped like flat plates or brackets, and vary from a few centimetres to a metre or more across; they are grey at first with a surrounding white margin, but then the slightly corrugated hymenium or spore-bearing surface develops and becomes covered in millions of rust-red spores, which are eventually liberated and cover the surroundings with red dust. As Dry rot infections in buildings are generally concealed, the appearance of a sphorophore may be the first sign of damage, although a characteristic mushroom smell may be detected if the infected building is closed for several days.

Dry rot can cause severe Brown rot damage with pronounced cuboidal cracking, the cubes being up to 50 mm along and across the grain. Decayed wood crumbles easily between the fingers to form a soft powder, a useful diagnostic feature, another being the brittle nature of the rhizomorphs when dry. The two most important features of Dry rot development are the fact that it usually develops in concealed timbers or on the concealed rear surfaces of skirtings or within the centres of beams, wherever the conditions are poorly ventilated, and the infection can also spread widely to dry wood in unventilated conditions through the ability of the rhizomorphs to convey water and nourishment. A related species *Merulius himantioides* is sometimes found causing similar decay in Scotland and other northern European countries.

The Cellar fungus *Coniophora puteana*, formerly known as *Coniophora cerebella*, is the most common cause of Wet rot in buildings and elsewhere where persistently damp conditions arise through soil moisture or plumbing leaks. Spores germinate readily and this fungus occurs inevitably whenever suitable conditions arise. The hyphae are initially white but growth is not as generous as for Dry rot. Mycelium does not develop, but thin rhizomorphs are formed which become brown and eventually black, although they never extend far from the wood. Sporophores are rare in buildings but consist of a thin skin on the wood covered with small irregular lumps. The hymenium, initially yellow, darkens to olive and then brown as the spores mature. Wood in contact with a source of moisture, such as damp brickwork, often consists of a thin surface film concealing extensive internal decay. The rotted wood is dark brown with dominant longitudinal cracks and infrequent cross-grain cracks. The easiest method for controlling this Wet rot in buildings is to isolate wood from the source of dampness.

Tapinella panuoides, formerly known as *Paxillus panuoides*, causes decay similar to that caused by the Cellar rot but tends to occur in much wetter conditions. The hyphae develop into fine branching strands which are yellowish but never become darker, and a rather fibrous yellowish mycelium, perhaps with lilac tints, may develop. The wood is stained bright yellow in the early stages but darkens to a deep reddish-brown with shallow cracking. The sporophore has no distinct stalk but is attached to the wood at a single point, tending to curl around the edges and eventually becoming rather irregular in shape. The branching gills on the upper surface radiate from the point of attachment. The colour is dingy yellow but darkens as the spores develop; the texture of the sporophore is soft and fleshy. In exceptionally damp conditions *Peziza vesiculosa* may develop, producing cup-shaped sporophores with an incurved margin, pale yellowish or buff coloured, sometimes wrinkled and often shiny. In buildings this fungus is usually found on wood affected by leaks from WC pans or soil pipes, although the fungus also occurs in cellars in old buildings where timbers such as door frames and stair strings are exceptionally wet.

The White Pore fungus *Antrodia sinuosa*, formerly known as *Poria vaporaria*, occurs occasionally in buildings. It requires a higher moisture content than Dry rot, but it is much more tolerant to occasional drying than the Cellar rot and is therefore normally the fungus associated with roof leaks. Growth is generally similar to that of Dry rot but remains white, whereas it becomes yellow or lilac for Dry rot and brown or black for Cellar rot. The rhizomorphs may be up to 3 mm in diameter, but are not so well developed as those of Dry rot and remain flexible when dry. When examined on the surface of a piece of wood or adjacent masonry, they appear to be distinctly flattened or having a flat margin on either side of the main strand, and they do not extend far from the wood. The development of a sporophore is rare in buildings but sometimes occurs in greenhouses or conservatories following severe decay; it is a white irregular plate 1.5–12 mm thick covered with distinct pores and sometimes with strands emerging from the margin. Wood decay is similar to that caused by Dry rot but the cubing is smaller and less deep; when decayed wood is crumbled between the fingers it is rather more fibrous or gritty. *Antrodia xantha*, formerly known as *Poria xantha*, is a similar Brown rot which is frequently found in greenhouses and conservatories, usually with no visible surface growth although mycelium may be found in cracks, particularly in the cubing cracks in severely decayed wood. A thin skin of yellowish white mycelium occasionally occurs. The sporophore is a thin yellowish layer of pores, distinctly lumpy when situated on a vertical surface. Another related species *Poria monticola* is also occasionally found in buildings, but only on softwood imported from America. This fungus is one of the most common causes of dote or pocket rot, in which isolated pockets of decay develop from infection that originated in the forest. Dote usually takes the form of small pockets of Brown rot with cracking along and across the grain. The pockets may be filled with fluffy white hyphae if the infection is actively developing. In the early stages only a brown stain may be present but rot can be confirmed by brashness or brittleness detected by lifting fibres with the point of a knife. Generally dote is halted naturally by the dry conditions in a building, although wet conditions can sometimes cause the development of extensive active decay. Some years ago the floors in large foodstores were generally Douglas fir (Columbian pine) imported from North America, and in stores with a daily floor washing policy dote development sometimes caused extensive decay damage. Other species can also cause dote or pocket rot in softwood imported from North America, such as *Trametes serialis* and *Lenzites trabea*; these additional species are named only so that an investigator has some guidance if referring for further information to specialist publications such as the author's book *Remedial Treatment of Buildings*.

The Stringy Oak rot *Donkiaporia expansa*, formerly known as *Phellinus cryptarum* or *P. megaloporous*, occurs in oak in conditions in which the Cellar or White Pore fungi are normally found in softwoods, such as in association with roof leaks or masonry dampened by soil moisture. As this fungus prefers oak or chestnut it is largely associated with older buildings constructed from these woods. It is a White rot, causing no distinct colour change in the decayed wood which becomes progressively softer with a longitudinal fibrous texture developing; the wood does not powder in the same way as wood decayed by Dry rot or the other Brown rots. Yellow or brown mycelium is sometimes formed on the surface of the wood. The sporophore is a thick tough plate or bracket, buff coloured but darkening as the spores develop. *Poria medulla-panis* causes very similar decay in oak, particularly in old timber-framed buildings where the oak is exposed to the weather, but there is no practical reason in terms of preservation why these two species should be differentiated.

Trametes versicolor, formerly known as *Coriolus* or *Polystictus versicolor*, is the commonest cause of White rot in hardwoods, especially in ground contact, but it is usually confined to light coloured tropical hardwoods, including the sapwood in durable species. The sporophore is rarely seen but consists of a thin bracket up to 75 mm across, grey and brown on top with concentric hairy zones and a cream layer underneath from which the spores are released. Infected wood initially suffers light flecking and eventually bleaches in colour. Shrinkage is rare and the decayed wood simply appears to be lighter and much weaker than sound wood. Although White rots tend to be associated with hardwoods in buildings, this species and less common White rots are sometimes found causing decay in sapwood of softwoods, particularly in external painted joinery such as window and door frames; this fungus most commonly occurs in tropical hardwoods used for window frames, particularly in sill sections but often spreading to the supporting softwood framework.

It will be appreciated from this brief description of wood-destroying fungi in buildings that they are invariably associated with a source of moisture. Visible defects are not usually tolerated and repairs are carried out before fungal infection can develop, so that extensive fungal decay is usually confined to concealed areas where it may be difficult to detect or where it develops through limited ventilation.

The Cellar rot usually occurs in floor plates in direct contact with damp masonry, usually causing progressive deterioration and slow settlement of the floor. Door posts and skirtings in contact with damp floors or walls are also frequently affected. However, in basements that are subject to periodic flooding, the conditions are usually too wet for this fungus and *Tapinella panuoides* may be found instead, particularly in the bases of basement door posts and the bottom of stair strings. If continuous dampness is replaced by periodic dampness in these conditions, there is a danger that Dry rot will develop during slow drying as the moisture content passes through the critical level for Dry rot spore germination; when Dry rot infection has developed in this way it will then spread extensively behind skirtings and plasterwork and through walls. In the roof of a building, periodic leaks in a relatively well ventilated area will usually result in the development of the White Pore fungus or, in oak and chestnut in older buildings, the Stringy Oak rot. Dry rot is comparatively rare in roof areas, except where unventilated conditions occur, usually in flat roofs, but it is particularly common in lintels, shutter boxes, sash boxes and other parts of window frames which are affected by dampness from, for example, a roof leak or faulty rainwater downpipe. Dry rot usually develops in a flat roof through interstitial condensation in boarding supporting the roof covering, the condensation progressively accumulating until it reaches the critical level for Dry rot development; interstitial condensation problems are described more fully in section 4.7.

There are various ways in which Dry rot can develop and cause serious damage in buildings, and it will probably be helpful to give several examples of actual incidents. Severe frost is comparatively rare in the Channel Islands, but sometimes high pressure weather systems settle over the British Isles which cause cold easterly winds over the Channel Islands with severe frosts. In unoccupied houses, or occupied houses with poor plumbing insulation, these frosts often cause fracture of the high pressure supply pipes to roof header tanks and severe flooding percolates downwards through the building, spreading more widely through each successive floor, basically affecting a cone-shaped zone. Although domestic insurance policies do not usually cover the cost of repairing leaks, they cover the rectification of consequential damage. Insurers are naturally keen to minimise claims for loss of use and usually promptly arrange for necessary repairs and redecoration, but often without considering the severe risk of subsequent Dry rot infection. All affected timbers that are subject to limited ventilation are at risk, including suspended floors, and extensive Dry rot damage can usually develop within two years of the original flooding. Most house owners consider the Dry rot development to be a separate incident, whereas it is actually a direct consequence of the original flooding and therefore covered by insurance.

In terrace buildings, or even in separate closely adjacent buildings in urban areas, dampness in one building may affect another, but Dry rot developing in one building is perfectly capable of spreading through masonry into another. Where Dry rot affects party walls it is therefore essential to carry out exposure work and treatment in both properties, however inconvenient this may be. In one tall building in London, Dry rot developed around the same part of the party wall at all levels above ground floor. Access was eventually obtained to the adjacent building which was occupied by a well-known financier, who was reputed to provide himself and his friends with suitable lady 'escorts'. The affected wall was adjacent to bathrooms on each level which served rooms occupied by these ladies who apparently cavorted rather energetically in the baths, causing severe wetting of the wall and extensive Dry rot development in timbers of intermediate moisture content, particularly in the adjacent building. This development of Dry rot through the establishment of a moisture content gradient between a source of moisture and dry conditions usually occurs in association with a plumbing leak, such as a defective connector or a floorboard nail through a heating pipe.

Dry rot does not occur only in older buildings but can affect new buildings if suitable conditions arise. A substantial commercial building in Ireland was designed without natural ventilation to conserve energy, ventilation normally being provided through the air conditioning system. Following completion a strike prevented immediate occupation of the building and the air conditioning was switched off for economy, the moisture of construction then causing the relative humidity in the building to rise towards saturation. The ground floor structure included concrete floor slabs isolated from the surrounding walls by proprietary fibreboard joints, with screeds and tile floors in some areas but chipboard floors on battens in areas where flexible service wiring was required; a proprietary floor system was involved in which the floor battens were isolated from the supporting concrete slabs by foam plastic. In this form of construction Dry rot always develops in the fibreboard movement joints, but in solid floors it is never seen as the joints are sealed with mastic. However, the joints are not sealed when a batten and chipboard floor is laid over the slab. The effect of switching off the air conditioning was to allow the relative humidity in the building to increase, approaching saturation beneath the chipboard floors, which enabled the Dry rot developing in the fibreboard joints to spread into the floors and cause extensive decay in the chipboard and supporting softwood battens.

6.5 Insect infestations

Furniture beetles comprising the sub-family Anobiidae are the best known wood borers in the British Isles, mainly because damage sometimes occurs in furniture and is thus readily apparent to a householder. The sub-family can be further divided into the Anobiinae which include the *Anobium* species, producing elongate ovoid or rod-shaped pellets, and the Ernobiinae which include *Ernobius* and *Xestobium* species, producing bun-shaped pellets, a useful diagnostic feature. All the larvae are curved and the life cycles tend to be long with very slow development of infestations over many years. All furniture beetles lay eggs in cracks or old flight holes; considerable damage may occur within the wood before the infestation becomes apparent through the emergence of adult beetles through flight holes and resulting boredust discharges.

The Common Furniture beetle *Anobium punctatum* is economically the most significant wood-borer, except in areas where there is a risk of damage by the House Longhorn beetle; the Common Furniture beetle does not usually cause as severe structural damage as this other wood-borer, but it is very widely distributed and an infestation invariably develops wherever suitable wood occurs. The Common Furniture beetle attacks the sapwood of both hardwoods and softwoods, extending into the heartwood in some light-coloured hardwoods or when damp conditions occur. The adult beetle is 2.5–5.0 mm long, the female usually being larger than the male and reddish to blackish brown with a fine covering of short yellow hairs over the thorax and elytra, particularly on freshly emerged insects; the elytra are the wing cases, whilst the thorax is the middle section of the body which is hooded over the downward projecting head in this particular beetle. Longitudinal rows of pits or punctata are readily visible on the elytra; they account for the Latin name *punctatum* for this beetle. When viewed from the side the hooded thorax is seen to have a distinct hump, another useful diagnostic feature.

The adult beetles usually emerge from the wood between late May and early August, and can be seen crawling over walls and windows. The beetles are strong fliers on warm days and live for 3–4 weeks, during which they mate and the female lays about 80 lemon-shaped white eggs, each about 0.3 mm long, in small groups in cracks, crevices, open joints and old flight holes. An egg hatches after 4–5 weeks, the larva breaking through the base of the egg and then tunnelling within the wood in the direction of the grain. The tunnels or galleries are filled with loosely packed gritty bore dust consisting of granular debris plus oval or cylindrical pellets, confirming that this beetle is an Anobiinae. When fully grown the curved larva is about 6 mm long with five-jointed legs, another diagnostic feature. The larva grows and eventually forms a pupal chamber near the surface about 6–8 weeks before emerging through a round flight hole about 1.5 mm in diameter. Under optimum conditions the life cycle can be as short as 1 year but it is usually longer and up to 4 years.

As infestation is usually confined to sapwood, damage is not structurally important, except where individual components in furniture are composed entirely of sapwood or where old types of blood or casein adhesives have been used which encourage development. Infestation is also encouraged by dampness, slight fungal or bacterial activity enabling the infestation to spread into normally resistant heartwood. These encouraging factors tend to result in larger beetles and shorter life cycles, particularly in stables and byres which are often extensively damaged. This damage is caused only by the larvae tunnelling invisibly within the wood, the only indication of current activity being bore dust discharge from new flight holes which suggests recent emergence, although vibration caused by an inspection can also dislodge bore dust. It is therefore often difficult to decide whether an infestation is currently active and remedial treatment is necessary or, if a treatment has been completed, whether it has been effective.

The Common Furniture beetle suffers very high natural mortality, perhaps only 60% reaching the larval stage when they may be further reduced in numbers by the action of predators. The most common predators are two Hymenoptera, the flying ants *Theocolax formiciformis* and *Spathius exarata* which can often be seen exploring flight holes. Predatory beetles are also sometimes found, such as *Opilio mollis* and *Korynetes coeruleus*, although the latter is more often associated with the Death Watch beetle. The names of these various insects are given here so that other reference works can be consulted if necessary to confirm identification; the common wood borers are shown in Figure 6.2 but the book *Remedial Treatment of Buildings* by the author contains an appendix with a key, which can be used for the identification of wood borers and other insects infesting buildings.

The Common Furniture beetle is certainly the most commonly encountered wood borer in the British Isles and, in fact, throughout all temperate parts of the world, but there are a number of other borers of varying importance. The related Anobiid beetle *Ptilinus pectinicornis* is sometimes found causing damage in hardwoods, such as beech, sycamore, maple and elm, and it can be a nuisance when it occurs in furniture. The damage is very similar to that caused by the Common Furniture beetle, including the size and shape of the flight holes. As treatment is the same it is unnecessary to distinguish between damage caused by the two borers, but it is important to be able to identify adult beetles that are found in buildings, in order to confirm that they are wood borers and that they have probably emerged from infested wood in the vicinity. This beetle is, in fact, very easy to identify as it is similar to the Common Furniture beetle in size and form, except that the thorax is not humped over the head and the antennae are comb-like in the male but serrated or saw-like in the female. Whilst this beetle is relatively easy to identify, another Anobiid known as the Drug Store, Biscuit or Bread beetle *Stegobium paniceum*, formerly known as *Sitodrepa panicea*, is far more difficult to distinguish from the Common Furniture beetle; it is found in woody natural drugs and cork, but most commonly infests biscuits, particularly dog biscuits, and the appearance of adult beetles may generate fears of serious woodworm infestation. It can be distinguished from the Common Furniture beetle as it lacks the distinctly humped thorax which is a feature of the Common Furniture beetle alone.

Another related Anobiid beetle is *Anobium pertinax* which is sometimes found in

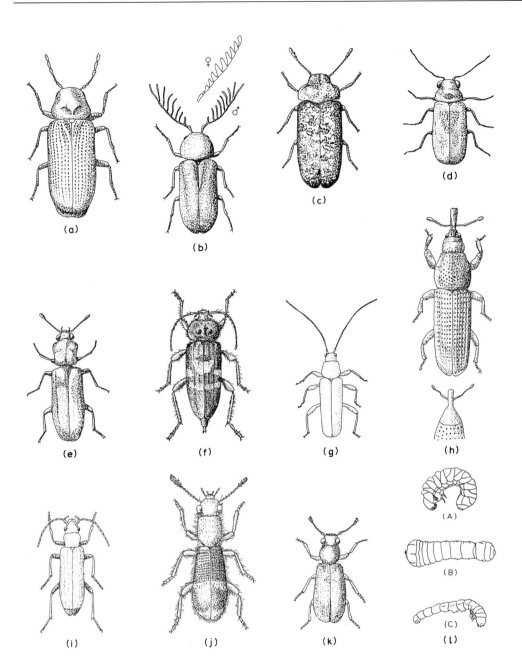

Fig. 6.2 (a–d) *Anobiidae* (furniture beetles) and (e–l) other common wood-boring beetles and their predators. (a) Common Furniture beetle (*Anobium punctatum*), (b) *Ptilinus pectinicornis*, (c) Death Watch beetle (*Xestobium rufovillosum*), (d) *Ernobius mollis*, (e) *Lyctus brunneus*, a Lyctid, (f) House Longhorn beetle (*Hylotrupes bajulus*), (g) Oak Longhorn beetle (*Phymatodes testaceus*), (h) Wood weevils (*Euophryum confine* and head of *Pentarthrum huttoni*), (i) Wharf borer (*Nacerdes melanura*), (j) *Paratillus carus* (a predator on Lyctid Powder Post beetles), (k) *Korynetes coeruleus* (a predator on the Death Watch beetle (*Xestobium rufovillosum*), (l) Typical wood-borer larvae: (A) Death Watch beetle (*Xestobium rufovillosum*), (B) House Longhorn beetle (*Hylotrupes bajulus*) and (C) Wharf borer (*Nacerdes melanura*).

the British Isles, usually in the North of England or Scotland, although it is much more common in Scandinavia. The damage caused by this beetle is very similar to that caused by the Common Furniture beetle but it is always associated with fungal attack, so that severe infestations are normally confined to poorly ventilated cellars or sub-floor spaces, or in wood subject to periodic rainwater or plumbing leaks. These same conditions encourage Common Furniture beetle activity and, as damage is so similar and there is no difference in treatment, it is really unnecessary to distinguish between these species.

All the beetles that have been mentioned so far are from the Anobiinae, which can be distinguished by elongated ovoid or rod-shaped pellets in the bore dust. There are also two significant beetles from the Ernobiinae which produce bun-shaped pellets. The Death Watch beetle *Xestobium rufovillosum* is the largest Anobiid beetle, and is always associated with incipient or active fungal decay. Infestations in the British Isles occur most commonly in oak in old buildings, but only because this wood was used extensively in construction, as infestations can also occur in chestnut, elm, walnut, alder and beech. Sapwood and heartwood can both be infested if previously infected by decay, and the infestation can spread into adjacent softwoods. Infestation is always confined to damp or decayed areas and a common characteristic of Death Watch beetle attack is the brown coloration of the infested wood due to Brown rot activity. It often appears that this beetle favours churches, but this is really a combination of circumstances which results in church timbers being particularly suitable; the roofing often consists of lead or other metal sheet and the periodic heating results in interstitial condensation within the supporting boards, causing the incipient decay which encourages infestation. The length of the life cycle depends upon the quantity of nitrogen available in the form of decayed wood and can be a single year under optimum conditions, although it is usually far longer and often as much as 10 years. The adults normally emerge between the end of March and the beginning of June but, where longer life cycles are involved, fully developed beetles may be found at other times just below the surface of the wood and apparently ready to emerge; the larva actually pupates in July or August, metamorphosing into an adult after only 2 or 3 weeks, but remaining in the wood and gradually darkening in colour until it emerges during the normal flight season the following spring.

The adult is not an active flyer and tends to mate with other beetles emerging from the same piece of wood, attracting their attention by bending the legs so that the head is struck on the wood surface producing a series of 8 to 11 taps in a period of about 2 seconds. Tapping with the tip of a pencil can generate a similar noise and can stimulate a response. This tapping noise probably accounts for the name of Death Watch beetle, perhaps because the sound is apparent in a house which is quiet through death. The tapping should not be confused with the sound produced by the Book louse *Trogium pulsatorium* which is more like a watch ticking than a tapping. Although the Death Watch beetle can fly reasonably well when the weather is very warm, it is probable that this pest is now spread largely by re-use of old infested wood. The Death Watch beetle has a much more limited distribution than the Common Furniture beetle in the British Isles, being confined to England, Wales and part of Southern Ireland. Death Watch beetle is common in the Channel Islands, but it is possible that a different strain is involved as heavy infestation is often found in softwood in the complete absence of any hardwood.

Ernobius mollis, sometimes known simply as the Bark borer, is an Ernobiinae which produces bun-shaped pellets similar to those of the Death Watch beetle, but it is completely different in all other respects. This borer produces galleries within and beneath softwood bark, sometimes extending to a depth of perhaps 10mm into sapwood. Only fresh softwood with waney edges (attached bark) is attacked and the infestation dies out naturally after 1 or 2 years, but the presence of galleries is often diagnosed incorrectly as Common Furniture beetle infestation and used as a justification for treatment. The concentration of activity beneath the bark and the galleries filled with bore dust containing bun-shaped pellets which are either brown or white in colour, depending upon whether the larva was feeding on bark or wood, enable the activity of this borer to be readily distinguished from that of the Common Furniture beetle with its deeper galleries and white oval or cylindrical pellets.

Wood infested by the Death Watch beetle may also be attacked by other borers, particularly the Common Furniture beetle and the Wood Weevils, with perhaps the rare *Helops coeruleus* in the damper zones. Predators may also be present, particularly *Korynetes coeruleus*, a distinctive blue beetle which is an active flyer and often the first sign that a Death Watch beetle infestation is present in concealed damp timbers. Of all these borers the Wood Weevils are most important, but they are not Anobiidae and behave quite differently.

Wood Weevils are Curculionidae which have a head formed into a narrow snout in front of the eyes. In the British Isles *Pentarthrum huttoni* is a native species usually found in buildings in decayed floorboards and panelling, as well as in old casein-glued plywood. *Caulotrupis aeneopiceus* is another native species more rarely found in buildings, and then only in association with very decayed wet wood in cellars and sub-floors spaces. Several other species may be found but, in recent years, the weevil that has attracted most attention is *Euophryum confine*, a species that was apparently introduced to Britain from New Zealand in about 1935. It has since spread very widely, apparently because it is able to infest wood which is not significantly decayed and which may have a moisture content as low as 20%. However, it is not really necessary to distinguish individual species or even to clearly distinguish damage from that of the Common Furniture beetle, except to note that, if infestation occurs in damp conditions or if active decay is present, these features may be of greater structural significance than the borer damage itself.

Whilst Common Furniture beetle infestations represent the most common borer damage in the sense that infestation may be encountered in any building within the British Isles, and Death Watch beetle infestation may be less known as it seems to attack the most valuable historic buildings, it is the House Longhorn beetle *Hylotrupes bajulus* that represents the most serious borer problem in the areas in which it occurs. It is largely confined to South-East England where it is sometimes known as the Camberley borer; other parts of the British Isles seem to be unsuitable for climatic reasons. This is a relatively large beetle which leaves a characteristic oval flight hole about 10 mm across. The appearance of even a single flight hole indicates that severe damage has already occurred and the condition of the structure should be checked by thorough probing. Because of the seriousness of the damage caused by this beetle the Building Regulations now require all structural wood to be preserved in the areas where this pest is established.

House Longhorn beetle attacks the sapwood of dry softwood. The adult beetle is somewhat flat and 10–20 mm long with slightly shorter antennae, the male being smaller than the female. The beetles are brown to black in colour, except that they have thick grey hairs on the head and the prothorax (the front section of the thorax). The female has a central black line and a black nodule on either side, whilst the male has white marks in place of the nodules. There are also distinct shaped white spots on the elytra (wing covers). The beetles emerge in July to September; a single female can lay as many as 200 eggs which hatch within 1 to 3 weeks, these eggs being spindle-shaped and about 2 mm long. In a roof the larvae from a single clutch of eggs can cause substantial damage within a period of 3 to 11 years before they pupate and emerge as adults, perhaps entirely destroying the sapwood and leaving only a thin surface veneer, slightly distorted by the presence of the oval galleries beneath. The first sign of damage may therefore be the collapse of a component consisting largely of softwood, although in warm weather the gnawing of the insects can be clearly heard. When fully grown the larva is about 30 mm long, straight bodied and distinctly segmented with a slight taper and very small legs. Pupation occurs in a chamber just beneath the surface and is complete in 3 weeks, the emerging beetle leaving the characteristic oval flight hole about 10 mm across.

Several other Cerambycidae or Longhorn beetles are sometimes found causing damage to building timbers. The two-toothed Longhorn beetle *Ambeodontus tristus* causes serious damage to softwoods in service in New Zealand. It thrives in similar conditions to the House Longhorn beetle and causes similar damage, but the oval exit holes are distinctly smaller at about 5 mm across. In 1974 this insect was found to have caused severe damage to joists in a cellar in Leicestershire. The infestation was introduced in the joists which were found to be *Dacrydium* pine imported from New Zealand. Whilst there appears to be no biological reason why this beetle could not become established in England, it must be emphasised that it is most unlikely and, if an oval flight hole of an appropriate size is discovered, it must first be suspected either that it was caused by a forest Longhorn beetle before the wood was converted or that it is a false flight hole caused by sawing timber through existing galleries of an older infestation.

In older buildings constructed from oak the sapwood is sometimes found to be attacked by the Oak Longhorn beetle *Phymatodes testaceus*. Eggs are laid in cracks in the bark and larvae bore between the bark and the wood, eventually making deeper tunnels to form pupation chambers. Damage is normally caused in the forest in sickly trees but can also occur in stored logs and boards being air-seasoned with waney edges. This beetle is sometimes found in buildings attacking dry oak but only if the bark remains; the damage is only superficial and the beetle can be eradicated by removing the bark, perhaps accompanied by a spray treatment to kill any surviving insects, although such a treatment must be mainly regarded as a precaution against Common Furniture beetle which is a far greater risk in oak sapwood. The Oak Longhorn beetle can also attack a number of other hardwood species, but if hardwoods are decayed other Longhorn beetles may be found, such as *Rhagium mordax* and *Leiopus nebulosus*. Whilst these beetles are capable of infesting damp decayed oak in buildings, such infestation is extremely rare unless it has been introduced to the building on the decayed wood involved; these Longhorn beetles are most commonly found in buildings on fire logs.

There are several other wood-boring beetles that are occasionally found in buildings. The most important are the Powder Post beetles which comprise the *Bostrychidae* and the *Lyctidae*. The European Bostrychidae adult beetles tunnel in bark in order to lay eggs, producing galleries which are free from dust. The hatching larvae then bore in the sapwood in search of starch, producing galleries which are packed with fine bore dust, this pattern of tunnelling enabling these insects to be distinguished from the Lyctid Powder Post beetles. Damage is principally confined to the sapwood of green hardwoods, although softwoods are occasionally found to be attacked, particularly if they have bark adhering. The European Bostrychidae are all small dark brown or black beetles, with the single exception of *Apate capucina* which has brown or red elytra and which is encountered occasionally in the sapwood of freshly felled oak. Bostrychid damage is not so common as Lyctid damage in buildings, probably because infestation commences with a tunnel bored by the adult which is readily visible, so that infested wood is rejected before being used, in contrast with Lyctid infestations which are initiated by egg laying and completely invisible.

The Lyctid beetles are all small with a length of only 4 mm, elongate but flattened in appearance with the head clearly protruding in front of the thorax. The colour of the various species encountered in Europe varies from mid-brown to black. The beetles are active flyers, particularly on warm nights. After mating the female lays 30 to 50 eggs in large pores on the end-grain of suitable hardwoods containing adequate starch; this means that only freshly converted wood can be attacked. The larvae develop progressively within the wood and pupation eventually occurs in a chamber just beneath the surface. The adult beetle normally emerges after 1 or 2 years in the spring, summer or autumn, usually between late May and early September, leaving a flight hole 0.8–1.5 mm in diameter. The galleries are packed with soft fine bore dust but are not distinctly separate, as are those of the Furniture beetles, and all the sapwood may be completely destroyed except for a surface veneer, accounting for the name of Powder Post beetle.

As the initial attack consists only of an egg layed in an open pore, it will be appreciated that the first sign of damage is often collapse or alternatively the appearance of a flight hole, in either case an indication of extensive damage within the wood. Susceptible large-pored hardwoods may be infested soon after conversion during air seasoning or storage and the insect may be introduced to a building within the wood, generally in furniture or decorative woodwork. Eradication treatment is unnecessary as the progressive loss of starch has usually rendered the wood unsuitable for infestation by the time the attack has been discovered. However, it is important to be able to identify Powder Post beetle infestations so that they are not confused with Common Furniture beetle which requires eradicant treatment.

Damage caused by Ambrosia beetles, the Scolytidae and the Platypodidae, is occasionally seen in buildings. The adult beetles bore in the bark of freshly felled green logs, producing galleries beneath the bark in which their eggs are laid. At the same time the adult beetles infect the galleries with the Ambrosia fungus which develops on the sap, providing nourishment for the larvae when they hatch. The larvae also burrow, usually forming side tunnels off the original adult gallery and eventually boring deeper in order to pupate. The gallery patterns under the bark are usually characteristic of the species responsible. As Ambrosia beetles are unable to infest wood that is dry or free from bark, they represent a problem in the forest rather than in a building or furniture, but the larvae sometimes bore quite deeply in the sapwood of tropical species, even penetrating the heartwood in light coloured woods, creating galleries that are clearly apparent as they are marked by dark staining caused by the Ambrosia fungus along the grain on either side of the hole. Wood damaged in this way may be noticed in furniture and may cause unnecessary concern, so that it is important to be able to distinguish between this damage and current active infestations caused by other beetles. Galleries formed by small Ambrosia beetles are referred to as pinholes, whereas large galleries are usually called shotholes.

The Carpenter ants *Camponotus herculeanus* and *C. ligniperda* have been causing increasing damage in buildings in Scandinavia in recent years. In nature these insects tunnel into old trees affected by interior decay in order to establish nests, but the borers have also been infesting timber buildings, usually those situated within or close to forests. It has been reported that Carpenter ants will attack only wood that has already decayed, but this has been discounted in recent years as *C. ligniperda* is certainly able to attack and utilise dry wood. Damage is typically internal, being an irregular cavity in soft decayed wood, but laminar and following the growth rings in sound wood. *C. vagus* is occasionally found in Southern Europe and *C. pensovanicus* in North America, together with other species. These Carpenter ants are being introduced occasionally into the British Isles in imported wood and there are already reports of isolated infestations.

There are many other insects found in buildings which are confused with wood borers; most of these insects can be identified using the key in *Remedial Treatment of Buildings* by the same author. Some of these insects, such as Spider beetles and Carpet beetles, cause damage to carpeting and furnishings but not to structural woodwork. The rather striking Wharf borer *Narcerdes melanura* must be mentioned, as it attacks wood which is reasonably sound and suffering only incipient decay, but it favours wood wetted by seawater, and it is sometimes found in great numbers in streets and in buildings in areas close to docks or sea defences. It is 6 to 12 mm long, elongate and red-brown with distinct black tips on the elytra and long antennae. It is sometimes confused with Longhorn beetles, as well as with the carnivorous Soldier beetle *Rhagonycha fulva*, which occurs commonly on summer flowers but which lacks the black tips on the elytra.

Very tiny beetles are sometimes found in buildings and can cause considerable concern if it is suspected that they are wood borers, but they are usually 'plaster beetles' which feed on mould on damp plaster and wallpaper; they are described in *Remedial Treatment of Buildings*.

Sometimes large galleries with a distinctive white lining are found in timbers in old buildings, often accompanied by extensive smaller galleries over the surface of the timber. The white lined galleries are caused by the Shipworm and the superficial damage by the Gribble, both of which are marine borers. The occurrence of this damage in buildings is simply an indication that flotsam has been used in construction!

6.6 Chemical damage

Chemical damage to wood in buildings is very rare and is always associated, in normal circumstances, with exposure to flue gases. The oxides of sulphur and nitrogen in the gases form acids with moisture, and these cause hydrolysis of wood cellulose and separation of the surface fibres. A typical result is shallow softening of the surface, often also accompanied by a white 'growth' over the surface, sometimes discoloured brown by the flue gases, which is often incorrectly diagnosed as fungal growth; it is actually separated fibres, effectively paper pulp!

In older buildings this damage usually affects roof timbers adjacent to leaking flues, although sometimes beams are actually built into the flues and affected in this way. Damage is most frequently seen in cottages, but it was also seen in the Chapter House roof at Lincoln Cathedral, apparently caused by flue gases from a slow burning coke stove that had been removed many years before. The same type of damage is also sometimes seen in modern buildings where there are leaks in fireplace and flue systems which allow combustion gases to spread through the wall cavities, typically affecting the ends of joists penetrating the inner leaf at first floor level and sometimes the adjacent first floor skirtings.

Some hydrolysis of wood cellulose occurs naturally when wood is exposed to the weather so that the surface acquires a soft furry texture. The damage is structurally insignificant, but wood affected in this way is difficult to paint or varnish and it is essential to remove the damaged surface with conscientious sanding before coating.

6.7 Wood preservation

Although the sapwood of most wood species is non-durable, it is possible to find woods with natural heartwood resistance to almost all destructive organisms, although physical properties, availability and cost may preclude selection in this way. In buildings, in particular, structural precautions to prevent wood becoming wet are sufficient to avoid the major hazards of fungal attack and associated insect infestation. The use of damp-proof courses to isolate floor joists from soil dampness, overhanging eaves and efficient rainwater disposal systems are examples of structural precautions which are significant in decay prevention. It is often claimed that painting wood, particularly external joinery, protects it from rainfall and thus preserves it against decay. In fact, minor imperfections or damage to the paint film will permit absorption of water whilst the remaining paint 'protection' will simply restrict evaporation and thus cause the dampness to accumulate, as described in section 6.3.

The main preservative action of paint is as a barrier, isolating the wood from attacking insects or fungi. Damage may allow access to the wood and protection is clearly more efficient if the wood is impregnated in depth rather than simply coated.

This is the origin of pressure treatment with coal tar distillates such as creosote, but all treatments of this type cause a fundamental change in the appearance of the wood; to the attacking organism the wood has the appearance of being a solid block of creosote or paint, and the wood has this same appearance to the human observer. Such changes are often aesthetically unacceptable and, if the treatment is also dirty, it may be unsuitable for use in many situations. The alternative is to abandon the barrier principle in favour of toxic action, low retentions of highly toxic compounds achieving preservation without significant alterations in the physical properties and appearance of the treated wood.

Pressure impregnation with water-borne preservatives, particularly copper-chromium-arsenic formulations, is extensively used for the preservation of structural timbers in buildings. The wood must be dried to a reasonably low moisture content to allow space within the wood for the treatment, but the treatment then introduces a large amount of water. These moisture content changes and subsequent drying often result in considerable distortion, so that these treatments are only suitable for sawn structural woodwork where cross-section dimensions are relatively unimportant; where cross-section dimensions are critical, for example in window frames, skirtings and floor boards, it is normal to use organic solvent treatments which avoid movement and distortion defects.

Water-borne copper-chromium-arsenic treatments give wood a distinctive green coloration, which confirms that treatment has been applied, but it is not so simple to check whether treatment is adequate. Treatment retentions can be checked by chemical analysis, but the usual reason for inadequate treatments is failure to achieve sufficient penetration. Pressure impregnation is complex and expensive, and only justified if it achieves deep penetration and the treated wood is then used in a situation in which deep penetration is essential. The main advantage of deep impregnation is that it avoids the danger of accidental damage of the protective envelope, although damage is usually caused by the development of checks or fissures through drying following treatment or moisture content changes in service. Structural softwood species that can be deeply impregnated are not readily available, and in the British Isles treatment is normally applied to European redwood or Scots pine in which the non-durable sapwood is readily impregnated and the resistant heartwood possesses moderate natural durability. European whitewood must never be treated in this way, as both sapwood and heartwood are susceptible to decay and both are resistant to impregnation so that only a superficial treatment can be achieved, even using very high pressures or special treatment processes that are claimed to be advantageous. Douglas fir or Columbian pine heartwood is moderately durable, but the non-durable sapwood cannot be impregnated, unless the wood is incised, this process uses blades or spikes to open up pathways into the sapwood which then allow tangential spread, although incising is usually used only on round poles and it is not used to a significant extent on sawnwood used in buildings. Even when treatment is applied to a suitable species, problems can still arise. On European redwood cross-cutting will expose heartwood end-grain which must be protected by applying preservative on site to all cut ends, holes or other worked areas, a necessity that seems to be beyond the comprehension of most site carpenters! If treatment is applied to wet wood, only superficial penetration will be achieved, even in readily impregnated sapwood. A cross-section sprayed with a reagent to detect the treatment will easily identify this fault, the treated zone forming a uniform shallow band around the outer edge of the wood instead of completely impregnating all sapwood areas. Inadequate penetration is caused in this way by a failure of plant operators to check the moisture content of wood before impregnation, and is the most usual single cause of failure of copper-chromium-arsenic treatment; this inadequate penetration permits decay to develop in inner untreated sapwood zones from which it then spreads to heartwood zones, effectively destroying the structural strength of the timber component.

These comments have been specific to decay development in situations in buildings where decay is probable, that is high-hazard situations. The most obvious examples are piles, poles and posts in which the wood is embedded in soil, but similar decay risks can arise in timber frame construction through interstitial condensation, as described in sections 4.7 and 6.3, as well as in components such as sole plates which may be affected in some circumstances by more direct sources of moisture. In contrast, decay is only possible, representing a moderate hazard situation, in other structural or carcassing timbers in buildings such as joists and rafters, the main risks being associated with built-in ends and with leaks due to roof or plumbing defects; the other main hazard in these timbers is Common Furniture beetle infestation, as well as House Longhorn beetle infestation in the geographical areas in which it presents a risk. These moderate-hazard risks can be avoided with a relatively superficial treatment which requires to have only modest resistance to leaching, although it must be non-volatile. Whilst pressure impregnation with water-borne salts may represent a reliable and relatively inexpensive form of treatment, the wood is often wet when delivered to site, causing wood-working difficulties such as tool corrosion, and the need to treat cut ends and other work surfaces is an expensive inconvenience. Treatment with organic solvent preservatives, such as impregnation by double vacuum, as used for window and door joinery, is perfectly suitable, although end-grain treatment is still necessary; with window joinery the double vacuum treatment is normally carried out after the components are fully moulded and cut to length in order to avoid this difficulty. The obvious alternative is to treat all components on site after cutting to size and completing all working; the most convenient form of treatment is immersion in a dip tank using organic solvent preservative.

There is, however, one other form of treatment which is particularly suitable for all wood in buildings. In borate diffusion treatment freshly felled green wood with a high moisture content is sawn and dipped in a concentrated borate solution, and then wrapped and stored for several weeks to allow the borate to diffuse into the moisture in the wood. This form of treatment can achieve complete penetration, even in European whitewood which cannot be impregnated by pressure methods. The wood can be cut and worked without exposing unprotected surfaces. The treatment is not completely resistant to leaching, but it is particularly suitable for use in buildings where even leaks do not usually involve excessive leaching; the treatment is even suitable for use for joinery as the external paint gives adequate protection against leaching.

Generally failures in preserved timber can be attributed to poor treatment control as the preservative systems in common use are extremely effective when properly applied. For example, organic solvent treatments applied to joinery wood by double vacuum sometimes cause problems with subsequent paint systems, typically causing staining but in extreme cases even cissing, where the paint refuses to form a uniform film over the surface. These problems are caused by excessive loadings of the organic solvent treatment in the wood. Impregnation treatments generally involve placing the wood in a treatment cylinder, followed by drawing a vacuum to remove air from the wood, flooding the cylinder with preservative and then releasing the vacuum so that the atmospheric pressure forces the preservative into the wood. In pressure impregnation a further pressure is applied to encourage deep penetration, but this extra pressure is usually unnecessary with low viscosity organic solvent preservatives. This type of impregnation cycle fills most of the open spaces or cells with preservative and is generally known as a full-cell process. If the initial vacuum is omitted, the preservative will be forced into the wood by compressing the trapped air and, when the pressure is eventually released, this trapped air will expand and remove much of the preservative in the cell spaces, resulting in a lower treatment retention, which is usually known as an empty-cell process. These basic processes are varied to suit particular preservatives and circumstances, a normal double-vacuum impregnation treatment with a low viscosity organic solvent preservative involving a limited initial vacuum, impregnation at atmospheric pressure and a final intense vacuum to remove excessive preservative from the wood. Where European whitewood or spruce is being treated, this typical cycle is usually varied by introducing a low pressure to encourage penetration at the impregnation stage, but spruce which has been ponded or has remained at a high moisture content for a protracted period will not have the usual high resistance to impregnation, but may be very permeable, this excessive permeability sometimes developing in zones rather than uniformly throughout the wood. This typical whitewood cycle does not allow for such permeable zones, and it is the excessive preservative retained in these permeable zones that usually causes the painting defects that have been described earlier.

Preservation problems are not confined to pre-treatment preservative systems but apply equally to remedial treatments applied *in situ* in buildings to eradicate existing fungal decay or borer infestations, or applied as a precaution where a hazard is considered to exist. It is best to use low viscosity solvent formulations for treatments of this type because they cling readily to wood surfaces when applied by low pressure spray, spreading uniformly and penetrating relatively deeply; the use of solvent systems, however, is now discouraged as it is believed that they are harmful to health, although the obvious fitness of retired employees of remedial treatment companies who applied solvent preservatives throughout their working lives suggests that these precautions are based on media hype rather than established risk. Solvent systems are usually designed to be sufficiently persistent to achieve required penetration, but sufficiently volatile to disperse reasonably rapidly, special co-solvents often being necessary to ensure that the preservative remains uniformly distributed throughout the penetrated zone rather than migrating to the surface with the evaporating solvent; proprietary systems vary widely in these respects, although they may contain the same biocides. However, the normal organic solvents are petroleum distillates, such as kerosenes and white spirits, which are highly flammable. The main fire danger is during and for several days after treatment when the solvent is evaporating, but the danger can be reduced by suitable precautions, such as isolating the electric wiring during treatment and allowing ventilation – although serious fires sometimes occur. The usual causes are spraying live electrical fittings where poor connections cause arcing and ignition, unprotected light bulbs bursting when accidentally sprayed, and auto-ignition following preservative spray of glass-fibre and similar insulation; when dispersed on some fibres the solvents can start to oxidise, the temperature within the insulating fibrous material increasing until ignition occurs. Such accidents can be easily avoided by following normal precautions, but there are always inexperienced or careless firms or individual operatives who either do not know of the essential precautions or choose to deliberately ignore them; firms which are members of the remedial treatment section of the British Wood Preserving and Damp-proofing Association operate to a strict code of practice and their work is subject to regular inspection.

These fire dangers have naturally prompted a search for safer preservative systems. Less flammable petroleum solvents with higher flash points are available, but they need to be selected with care. The ideal system is a normal kerosene solvent with the lighter fractions removed to increase the flash point, as the volatility and drying time is not greatly affected by this change, suitable solvents being the high flash white spirits which are readily available throughout Europe, although quite difficult to obtain in the British Isles. Many preservative manufacturers have therefore used odourless kerosene in the mistaken belief that it also reduced the odour of the treatment; the word 'odourless' actually refers to the burning properties of the kerosene and not the smell when sprayed onto wood. One of the features of typical odourless kerosene is a 'tail', or residue of low volatility, which has quite a pungent odour if the preservative is applied excessively generously, but because of its low volatility this tail will be slowly released over a period of many months and it can cause the formation of extremely pungent odours when partly burnt. The most serious problems arise through partial combustion through the use of a gas cooker or a flueless gas or kerosene heater. In addition, this slowly released 'tail' can be absorbed into the plasticiser in PVC wiring, diluting the plasticiser and causing it to 'bleed' from the insulation, causing staining around roses on ceilings and from switches and sockets on walls. If the preservative has been applied particularly generously, the plasticiser solution will penetrate into light fittings, eventually causing arcing and the development of tracking across insulation which will cause fuses to blow or circuit breakers to trip. Electricians subsequently investigating these faults often diagnose failure of the PVC wiring, but the short circuits are actually restricted to phenolic insulators in fittings, and only the affected fittings need replacement.

Fire danger can be completely avoided by using water-borne formulations, usually solutions of water soluble biocides such as borates, or emulsions or suspensions of organic biocides, often in the form of concentrated organic solutions. Unfortunately water-borne formulations have several distinct disadvantages. If they are sprayed on timbers above the ceiling, accidental excessive application can result in serious staining. Generous application to floor timbers can result in cross-grain expansion and distortion, followed by excessive shrinkage on subsequent drying. Penetration is always much less than for low viscosity organic solvent systems, particularly with emulsion and suspension systems which tend to break at the surface, resulting in only superficial deposits of the biocides; emulsions of concentrated high flash organic solvent solutions are much more effective than simple emulsion or suspension systems, so that there are distinct differences between proprietary products.

Some of the phenolic preservatives, particularly the chlorophenols, can cause treatment operatives severe respiratory and dermal irritation, particularly if excessive spray pressures are used which result in preservative atomisation and spray drift. Dermal irritation problems are often due to the solvents alone, and enquiries usually disclose that the individuals are sensitive to similar solvents such as gasoline, kerosene, white spirit and turpentine; such problems are usually associated with fair skin and are aggravated by exposure to sunlight. This dermal irritation is also aggravated by some preservative biocides, particularly chlorophenols such as pentachlorophenol (PCP) and organotin compounds such as tri-*n*-butyltin oxide (TBTO); sensitivity to these biocides varies enormously and, if normal precautions are observed in use, problems are only encountered by particularly sensitive individuals. In extreme cases, respiratory irritation can occur and cause coughing and bleeding from the nose, but such reactions are usually related to extreme exposure, particularly spraying preservatives in roof spaces during very hot weather; these problems can be reduced by using a coarse low pressure spray to flood the surface of the wood with preservative which can then be absorbed by capillarity, thus avoiding high pressure sprays which cause atomisation and rapid volatilisation, but improved ventilation may also be necessary.

If sufficient ventilation is provided following treatment, the solvents will rapidly disperse, leaving only preservative deposits of low volatility which do not normally cause persistent odours. There have been various suggestions in recent years that these treatments are dangerous to health, but all treatments are approved by the Health and Safety Executive, and most preservatives were previously controlled under the voluntary Pesticides Safety Precautions Scheme operated by the Executive before the present mandatory system was introduced. The health risks associated with current preservatives have therefore been carefully assessed and there is no reason to suppose that they present significant risks, either to treatment operatives or to persons resident in treated buildings. Obviously treatment operatives are severely exposed and would be expected to suffer most seriously from any health hazards but, although there are perhaps 5000 to 10000 operatives employed in remedial timber treatment in the British Isles, reports of health problems are very few indeed, despite the fact that most operatives work within the industry for many years; on the contrary, it seems that operatives suffer less from some common illnesses such as colds and influenza!

There were several proprietary remedial treatment preservatives some years ago which were based on o-dichlorobenzene or on mono- or dichloronaphthalene. These biocides are oils which can be readily absorbed through the skin, and they certainly presented a danger of liver damage to treatment operatives. The dangers were much less with the solid polychloronaphthalene waxes which could not be absorbed in this way, and no illnesses were reported with these waxes, despite their extensive use at very high concentrations over many years. Pentachlorophenol attracted attention in the past because of its pungent and irritating odour when applied, but in recent years attention has concentrated on the dioxin impurities that may be present in chlorophenols. Many different dioxins can be formed, depending on the chlorophenols involved and their method of manufacture, but only a few of these dioxins are excessively toxic, particularly the dioxin associated with the herbicide trichlorophenyl acetate (2,4,5:T) which caused such concern during the Vietnam war; this dioxin is not associated with the tri-, tetra- or pentachlorophenols used in wood preservation which have a different structure and produce a different series of dioxins which are generally much less toxic than the chlorophenols from which they are derived. If dioxin impurities are present in a chlorophenol wood preservative, they are generally insoluble and precipitate as sludge at the bottom of the mixing or storage tanks, rarely occurring in normal products.

The chlorinated hydrocarbon insecticides DDT, Dieldrin and Lindane have attracted particular attention at times because of their interference in environmental food chains when used in agriculture or horticulture. Wood preservation treatments do not normally interfere in the environment in this way, and the use of Lindane is still permitted. In recent years much safer pyrethroid insecticides have been introduced and are being progressively used more widely, particularly Permethrin; various natural and synthetic pyrethroids have been approved for many years for use in the food industries for insect control.

The organotin compound tri-n-butyltin oxide has been in use in remedial wood preservation for about thirty years without any serious health problems being reported, other than the dermal and respiratory irritation previously described, although other organotin compounds are exceedingly toxic. This is a further example of the situation that has been encountered in relation to chlorophenol dioxins; although there may be some exceedingly toxic compounds within a group, that does not mean that all compounds in that group are similarly toxic and some may be virtually non-toxic. Other metal compounds, particularly copper and zinc naphthenates, have been used for very many years in the wood preservation industry without reports of unusual health risks, although the naphthenic acid liberated from these treatments has a distinct musty pungent odour which is unpleasant and irritating. A leading manufacturer of preservatives of this type has replaced the napthenates in recent years with acypetacs compounds which avoid this musty odour, although they produce instead a slight sickly odour to which some individuals seem to be particularly sensitive and which has caused difficulty in houses when excessive preservative has been applied.

Health problems are not confined, of course, to remedial treatment wood preservation, but obviously spraying in confined spaces represents the most intensive exposure that is likely to be encountered. The other common problems that arise

Wood problems

are dermal and respiratory problems due to handling timber treated with organic solvent preservatives, particularly in hot weather, and careless operation of organic solvent treatment plants involving preservative spillage, vacuum pumps discharging into working spaces, and timber dried after treatment in working spaces, all problems that are associated with careless handling rather than any defect in the preservative system. However, there are periods when a series of complaints arise, apparently associated with a particular preservative biocide. Obviously reports of problems can prompt further unjustifiable complaints, but a series of incidents usually have some common cause which is often very difficult to identify. Some of the complaints in recent years seem to be associated with Lindane treatments, whilst other complaints are associated with TBTO treatments. In both cases unusual volatility seems to be involved which sometimes affects treatment operatives, but which is also readily apparent to the occupiers of treated buildings. In such cases, it must be suspected that the Lindane and TBTO were poor quality products containing impurities which have caused these problems, as such problems are not associated with the pure compounds; Lindane is defined as 99% pure gamma isomer of hexachlorocyclohexane, previously known as gamma-benzenehexachloride, and it seems that some imported material contains much higher concentrations of other isomers. TBTO often contains a stabiliser, and it seems that some of the stabiliser compounds interfere with the fixation of the TBTO to the treated wood so that the TBTO remains volatile. Such problems may be indications of inadequate quality control but they are very rare; when complaints are investigated it is generally discovered that normal precautions have not been observed, particularly in relation to excessive treatment levels and ventilation following treatment.

Masonry problems

7

7.1 Introduction

Masonry comprises the parts of a building, mainly the walls, which are constructed using blocks of materials set in mortar, and includes stonework, brickwork and blockwork. Masonry suffers some structural movement due to moisture content and temperature changes, as described in sections 2.3 and 3.5, and it may be necessary to incorporate vertical movement joints to accommodate horizontal movement. Where the masonry forms an external weather protection to a structure of a different nature such as a timber, concrete or steel frame, horizontal movement joints may also be necessary to accommodate differential vertical movement.

Masonry deterioration is always associated with excessive moisture contents and design precautions are therefore necessary to minimise moisture absorption. Damp-proof courses must be provided to prevent absorption from the soil by capillarity, as well as copings, eaves and gutter details designed to minimise absorption at wall-heads, although such precautions cannot give protection against driving rain. It is therefore necessary when designing masonry to take account of the driving rain index for the area involved, adjusted to take account of local exposure conditions, as indicated in Building Research Establishment Digest 127. Special design precautions may be necessary if the driving rain index is high, particularly if severe frost or urban pollution is also involved, these design precautions involving detailing and selection of both the masonry material and the jointing mortar. Defects and deterioration in masonry can generally be attributed to failure to observe these design criteria properly.

A warning is necessary at this stage in relation to historic buildings. Severe masonry deterioration is often observed and reported, prompting a financial crisis in relation to repairs that are considered to be urgently necessary. Such developments can always be attributed to the new appointment of a person with critical responsibility for maintenance, such as a new architect or surveyor, or even a new administrator such as the Dean in the case of a Cathedral. Often careful study will disclose that the observed damage has developed over a period of many years, perhaps even over several centuries, and that the rate of deterioration is actually very slow. Such studies also often disclose that the most severe deterioration is associated with more recent repairs, often carried out using a stone which was selected for its good durability reputation whereas it has actually performed very poorly in relation to the original stone structure. This scenario is particularly apparent on the Cathedrals of the British Isles. For example, York Minster is constructed from magnesian limestone, which generally has a poor durability reputation, particularly in urban situations. Masonry restoration over a period of many years involved extensive replacement of deteriorated magnesian limestone with supposedly more durable calcareous limestone, but this restoration policy was disastrous. The different texture and porosity of the calcareous limestone became particularly apparent as severe darkening during wet weather, but colonisation by algae and lichens also developed, causing darkening even in dry weather. Progressive erosion of the calcareous limestone then developed, despite its excellent durability reputation. In fact, the durability reputation was based on performance in structures consisting only of a single calcareous limestone, and the deterioration was caused at York Minster by the calcareous limestone inserts receiving washings from adjacent magnesian limestone, the erosion being caused by the formation of magnesium salt crystals in the calcareous limestone. Similar compatibility problems arise when limestone washings are absorbed by some sandstones, and even when lime mortar washings are absorbed by adjacent sandstones, causing deterioration immediately below each horizontal mortar joint.

Frost and chemical damage in masonry is considered in detail in sections 4.3 and 4.4 in the chapter on moisture problems in buildings. Section 4.5 refers to metal corrosion, including problems associated with reinforcements and fixings in masonry; iron dowels are a particularly serious problem in Victorian masonry with rusting causing expansion and often severe fracture damage in critical parts of a structure, such as window mullions and pinnacles. The purpose of this chapter is therefore to describe the various masonry problems that may arise, but without repeating detailed scientific explanations that are available elsewhere in this book.

7.2 Stone masonry

Natural stone can be conveniently classified according to its origin into primary or igneous rocks, secondary or sedimentary rocks, and metamorphic rocks. The igneous rocks, which are formed by solidification of molten material, include the basic rocks, such as basalt and dolerite, often known as whinstone, and the acid rocks, such as granite. Whilst whinstone is most commonly crushed for road stone, granite is widely used as a building material but it varies greatly. Granite is composed largely of silica crystals which may be small, granular and open-textured, with an appearance of a rather coarse sandstone or large, compact and virtually impermeable to water.

The sedimentary rocks are formed from debris deposited by wind or water. In sandstones deposits of silica granules, usually from granitic rocks, are consolidated with an amorphous silica or calcium carbonate cementing matrix, often in conjunction with aluminium or iron oxides, the latter contributing yellow, orange or brown coloration. The properties of sandstones depend largely on the cementing matrix as the silica granules are virtually inert. The limestones are formed either from accumulations of animal shells in a calcium carbonate cementing matrix, or by crystallisation from solution to give a characteristic oolitic structure. Sedimentary stones of intermediate structure also occur and are termed arenaceous or siliceous limestones or calcareous sandstones, depending on whether silica or calcium carbonate predominate.

The metamorphic rocks developed from all these types under the influence of heat and pressure. Thus the schists and gneisses were formed from the igneous rocks, quartzite from sandstones, marbles from limestones, and slates from mud and shale.

Suitability for use in building depends on a number of factors, the most important being availability and workability. Obviously local stones have been preferred, but they have not always been ideal and the progressive improvement in communications over the years has allowed building stones to be obtained from ever widening areas. Many older buildings are constructed in solid masonry using porous sandstones or limestones. In this form of construction resistance to rain penetration depends on the balance between the water absorbed, the capacity of the wall, and the rate of subsequent evaporation, as explained in more detail in section 4.8. Heavy showers alternating with bright and windy periods are less likely to lead to major water accumulations than continuous drizzle with exceptionally humid conditions which obstruct evaporation, but almost always dampness will be apparent on areas of thinner wall section, such as at window reveals. Such problems are usually avoided in modern buildings by the use of cavity construction, but water penetration problems can still occur if the external masonry is exceptionally permeable, as significant water accumulations can then occur in cavities, particularly if weep holes are omitted at cavity trays over openings and at damp-proof courses, and there are sometimes problems around openings if the cavity closures are not properly protected against damp penetration and allow rain penetration.

Frost damage, described in more detail in section 4.3, is mainly associated with stone which is saturated with water during freezing conditions. A stone which has a high saturation coefficient, that is the amount of water absorbed by the stone is high in relation to total porosity, will tend to be most susceptible to frost damage, but the pore size is also very important. A microporous stone with small pores will tend to absorb water more powerfully by capillarity and release it more reluctantly compared with a macroporous stone with large pores, so that a microporous stone is more likely to be saturated during freezing conditions and thus more likely to suffer frost damage. This pore size factor is often recognised by experienced masons and quarry masters, who consider that a stone with coarser texture is likely to be more durable as it is 'able to breathe', a reasonable interpretation of the evaporation of trapped water that can occur from a macroporous stone.

Simple leaching damage is usually insignificant, except in the case of the more soluble magnesium carbonate limestones, but rainfall normally consists of carbonic acid, formed by the absorption of atmospheric carbon dioxide, and this weak acid attacks the carbonates in natural stone to form the more soluble bicarbonates. In polluted urban and industrial areas the much stronger sulphur and nitrogen acids may be formed in rainfall, greatly increasing the rate of carbonate erosion and even causing slow deterioration of silica. The rate of deterioration depends on the nature of the stone in terms of crystallite form and size, and pore volume and size. The relationship governing the rate of leaching appears to be complex but actually obeys the well-known natural law of mass action; leaching is greatest for stone with a high total porosity coupled with a small crytallite size. Thus an impervious carbonate, such as a marble, will suffer only surface etching, whereas a stone with high total porosity comprising pores of small diameter, that is a microporous stone, will tend to erode much more rapidly. In fact, this rapid erosion is due in part to the way in which water is retained within a microporous stone, permitting protracted absorption of atmospheric pollutants and thus prolonged acid attack of the stone components, but stone deterioration does not depend on acid leaching alone. This acid attack results in the formation of soluble salts, sulphate or nitrates if sulphur or nitrogen oxides are present as atmospheric pollutants. These salts accumulate at stone surfaces through water evaporation; crystallisation of these salts involves crystal growth which disrupts and erodes the stone surface, the rate of erosion depending on the amount of water of crystallisation involved, which depends in turn on the nature of the salt, and thus the nature of the stone and pollutants from which it is derived. These chemical deterioration processes, and similar frost deterioration processes, are dependant on the presence of moisture and are therefore described in more detail in sections 4.3 and 4.4.

Stone deterioration takes many different forms. The Portland limestone used for the construction of St. Paul's Cathedral in London is very durable, even in this severely polluted urban situation. On the upper surfaces of the copings, string courses and statuary where the stone is freely exposed to rainfall, the surface has slowly eroded through acid leaching and salt crystallisation within the surface pores. This erosion is very slow, typically removing 5–10 mm of the surface every 100 years. Erosion also occurs on vertical surfaces subject to water runs and driving rain; although this erosion is much less than that occurring on upper surfaces it is still sufficient to keep the surfaces clean. In contrast, areas beneath overhangs such as copings, cornices and string courses have become black. It was always thought that this was due to adhesion of smoke particles, but much of the darkening originates through biological colonisation, particularly small lichens; see section 7.6. The areas are also sufficiently damp for salt formation, but the salts are not washed away from the surface and calcium sulphate eventually seals the surface pores to form a dense surface crust. This crust is formed from material extracted by acid rain from the zone immediately beneath it, and eventually this weakening allows the crust to break away, usually prompted by some other event such as severe frost following a period of particularly wet weather.

It is often recommended that major limestone buildings in urban situations should be washed regularly to prevent the accumulation of a heavy calcium sulphate crust, but only prolonged washing is effective because of the limited solubility of the calcium sulphate; the removal of the crust tends to allow deeper penetration of acid polluted rainwater, probably exaggerating deterioration in the long term. A more sensible approach might be regular treatment of the stone surface with a persistent bactericide, such as the borate treatment mentioned in section 7.7, to control the sulphating and nitrating bacteria which are responsible for converting the sulphurous and nitrous acids formed from atmospheric pollutants to much more erosive sulphuric and nitric acids.

The type of damage caused by sulphate formation varies widely. Calcareous sandstones, that is sandstones in which silica particles are cemented by calcium carbonate, have been extensively used for the construction of major buildings in Edinburgh and Glasgow. These stones are much more porous than Portland limestone and polluted rainfall therefore penetrates more deeply, eventually resulting in the formation of a calcium sulphate densified layer typically 10–20 mm thick with a weakened layer beneath. The first sign of serious damage is usually distortion as the surface crust begins to separate from the stone, but freezing following rainfall dislodges pieces of the crust which separate or spall away. The susceptibility to damage and the depth of spalling vary between individual stones, and the building eventually acquires a very patchy and neglected appearance. In heavily polluted atmospheres, particularly in areas where coal fires are still extensively used, the calcium sulphate traps particles of carbon from the smoke which, with algal and lichen growth, eventually result in a black surface, any spalled areas then showing up as light patches. With more durable sandstone in which the cementing matrix is largely silica, spalling may be avoided but the black coating still develops. These forms of deterioration are particularly common in northern cities such as Leeds, Glasgow and Edinburgh where most of the buildings are constructed from sandstone. The traditional Edinburgh stone Craigleith resists spalling damage but acquires a black coating; this is so uniform that it is sometimes suggested that the architects selected this stone for the eventual black appearance of the buildings, but the black colour develops only very slowly and architects are always keen to see their buildings in pristine condition at the time when construction is completed so that they would have used a dark stone if they required a dark appearance. Perhaps the argument in favour of black buildings is simply an excuse to avoid the expense of cleaning! There are many other sandstones used in Edinburgh and Glasgow, most of them derived from local quarries, some of the Glasgow stones coming from railway excavations, the stones varying greatly in their colour, texture and durability.

At York Minster, another heavily polluted site, the form of deterioration is very different. The magnesian limestone involved is very dense and polluted rainfall penetrates to only a limited depth. A surface crust of calcium sulphate mixed with magnesium sulphate develops but it is very thin and peels away at intervals. Again the deterioration varies between stones; much of the Minster is constructed from very dense and very durable Huddleston which suffers only slow surface erosion, but Tadcaster magnesian limestone is also used which suffers particularly from the surface peeling effect.

7.3 Brickwork and blockwork

Bricks and blocks are simply synthetic stones, physical size usually deciding whether they are described as bricks or blocks. A brick is intended to be the optimum size to be held in one hand, whilst the other hand is used for wielding a trowel and applying mortar. Blocks are much larger and are usually placed using two hands following application of the mortar bed. Both systems have advantages and disadvantages, but generally brickwork is more uniform, particularly in regard to mortar continuity, and unprotected brickwork is more suitable for exterior use than unprotected blockwork.

Bricks are commonly manufactured by firing natural clay, although they are also manufactured from aggregate; calcium silicate bricks are manufactured by autoclaving sand or crushed flint with lime, but ordinary concrete mixes are used in some areas. The resistance of these bricks to frost and salt crystallisation damage depends, as for natural stone, on their porosity characteristics, macroporosity always being more durable than microporosity as explained in sections 7.2, 4.3 and 4.4. However, each type of brick has its own distinctive properties.

Clay bricks sometimes exhibit dark specks or brown streaks during weathering which are generally due to coal incorporated into the clay during manufacture. These effects are largely aesthetic, but natural clays also contain salts and shells which can cause more serious structural damage. Calcium, sodium, potassium and magnesium sulphates commonly occur and, in more permeable bricks, the more soluble of these sulphates can be leached into the adjacent mortar. This results in the single most serious problem with brickwork, as the sulphates can then react with the tricalcium aluminate content in ordinary Portland cement to cause sulphate attack, as described more fully in sections 4.4 and 8.6. In extreme cases sulphate attack can cause 10% expansion in the mortar, resulting typically in 2% vertical and 1% horizontal expansion in brickwork. Sulphate attack can only occur whilst the mortar remains alkaline, carbonation eventually resulting in resistance to further attack as well as loss of cohesion in affected mortar and often severe loss of structural stability. Sulphate attack will only occur in this way in brickwork if there is sufficient soluble sulphate present, the bricks are sufficiently permeable, and the use, situation and detailing results in sufficient exposure to rainfall for the sulphates to leach into the mortar. Susceptibility to sulphate attack is therefore partly due to brick type and partly due to structural design, particularly in free standing walls and parapets in which coping details are critical. The erection of brickwork in warm wet weather and failure to provide temporary protection can result in the development of severe sulphate attack over a period of only a few days; in other situations it may take 10 years or more before significant sulphate attack develops, and by that time the extent of the damage is usually reduced by the extent of natural resistance that has developed in the mortar due to carbonation. Within individual buildings the distribution of sulphate attack depends on the exposure of individual walls, so that it is least on walls protected by projecting eaves and most severe on unprotected walls, such as gables with flush verges, as well as being most severe on walls exposed to prevailing driving rainfall. Whilst the danger of sulphate attack can be completely avoided by careful choice of facing bricks, particularly by ensuring a low soluble salt content, indicated by designation L in British Standard BS 3921:1985 *Specification for clay bricks*, this precaution may not always be possible. Design details will reduce the danger of sulphate attack but cannot avoid the danger with, for example, exposed areas of walling such as gables facing prevailing driving rainfall. If designation L bricks cannot be used, the use of sulphate resisting cement should be a normal prudent precaution, whatever the exposure conditions.

Although brick manufacturers are very reluctant to admit that their products contain sulphate to a sufficient extent to induce sulphate attack in brickwork mortar, there have been a number of major claims in this respect in recent years and it has been necessary for brick manufacturers to be more careful in the way that they promote their products. For example, the London Brick Company Limited introduced *Constructional notes for areas of severe exposure* in October 1982 which, whilst it does not refer to sulphate attack of mortar due to sulphates in their bricks, emphasises the need to select bricks with care and to use sulphate resisting cement in exposed conditions. These notes were revised as *Climate and Brickwork; constructional notes* in October 1988. These and similar notes imply that sulphate attack is only likely in severe exposure conditions but, whilst it is true that sulphate attack is almost inevitable when some bricks are used in ordinary Portland cement mortar in severe exposure conditions, it is totally wrong to assume that sulphate attack cannot occur in less severely exposed conditions. Sulphate resisting cement should therefore always be used with facing bricks which contain too much soluble sulphate to be classified as designation L in BS 3921:1985.

Sulphate attack is not the only problem associated with soluble salts in clay bricks. If salts are leached onto the brickwork surface during rainfall they will subsequently crystallise as efflorescence, which can completely obscure the brick surfaces in extreme cases. Crystallisation of these salts can cause deep erosion on underfired bricks, and BS 3921:1985 therefore classifies susceptibility to efflorescence. Unfortunately BS 3921:1985 does not define requirements but only test methods; the earlier BS 3921:1974 required facing and common bricks of ordinary and special quality to have a liability to efflorescence of no more than *moderate*, and for special quality the soluble salt content was also defined as the same levels as Designation L in the current 1985 revision. Rather strangely, performance requirements are also currently omitted from the related British Standard BS 628:Part 3: 1985 *Code of Practice for use of masonry, Part 3, Materials and components, design and workmanship*, so that there are actually no current requirements for brick quality! However, there are the normal legal requirements that the bricks should be reasonably suitable for the purpose intended; in the case of facing bricks this must mean that they must be reasonably durable, they must be reasonably free from salt which may have a deleterious effect upon the mortar, and they must be reasonably free from efflorescence which would have a deleterious effect upon their appearance.

If brick clay contains sea shells, firing converts them to quicklime. Water is subsequently absorbed, forming slaked lime and causing expansion. This results in small pieces of brick spalling away from the surface, usually leaving a shallow circular cavity with the offending small shell at its centre, although sometimes larger shells occur which cause more serious irregular spalling. This damage is most severe when slaking has occurred slowly through the absorption of moisture from the atmosphere, so that if this lime blowing damage is observed on bricks before laying or during the early stages of laying, the damage can be greatly reduced by wetting the bricks to induce rapid less troublesome slaking. Although lime blowing can be a serious problem with some bricks when clay is being used which contains significant numbers of shells, it is not a defect that is extensively described in the technical literature.

There are two problems typically associated with calcium silicate bricks. The durability of the bricks depends on the grading of the aggregate, bricks manufactured from graded crushed flint being usually more durable than those manufactured from natural sand, and it is necessary to use only bricks with adequate durability classifications for external situations. The second problem concerns the relatively high movement with changes in the moisture content, typically 0.01–0.04% from wet to dry. This movement is very high compared with clay bricks, and it is essential to avoid the use of excessively strong mortars to ensure that the brickwork is able to tolerate this movement. Vertical movement in calcium silicate brickwork is not apparent, provided that the roof load is supported on the inner skin and there is a sufficient movement gap to allow for the movement in the external brickwork; however, horizontal movement often results in vertical cracking, usually through openings which represent the weakest zones of a wall, unless appropriate precautions are taken, particularly the use of mortar that is not too strong, but also the provision of vertical movement joints in long brickwork runs. The provision of vertical movement joints in this way is not, however, a special requirement for calcium silicate brickwork alone, as such joints are required also for clay and concrete bricks to allow for thermal movement.

Blocks are usually manufactured from various types of concrete. Normal dense blocks use normal concrete mixes based on natural or crushed aggregates, whilst lightweight blocks use either lightweight aggregates such as clinker, expanded clay or foamed slag, or alternatively use a fine powder aggregate coupled with aeration to reduce the density. Compressive strength varies enormously, although generally dense blocks are stronger than lightweight blocks. Thermal and moisture movement precautions in blockwork are the same as for brickwork, and blockwork is basically identical in properties to brickwork constructed from concrete bricks, except that jointing in brickwork tends to be more consistent and reliable than in blockwork due to the manner of construction. In some areas blockwork is used almost exclusively, but generally finished externally with render. Usually, external walls are constructed using independent blockwork leafs, joined only by normal ties as for brickwork walls, but in some areas hollow concrete blocks are used to construct solid walls, introducing thermal insulation problems, as well as water penetration problems unless great care is taken with the design and execution of the external render, as explained in more detail in section 7.4.

7.4 Mortar and render

Joints in brickwork and blockwork, as well as render and screed finishes, usually involve mortar, that is concrete prepared using only fine aggregate or sand. The strength of a mortar is very important in relation to its use, and mortar designations have now been generally adopted to classify mortars in this way, as shown in Table 7.1.

Lower mortar mix numbers indicate increased strength and improved durability, whilst higher mortar mix numbers indicate increased flexibility and improved

Table 7.1 Mortar designations

Mortar mixes (parts by volume)	Cement/lime/sand (dry hydrated lime)	Cement/sand (with plasticiser)
i	$1:0-\frac{1}{4}:3$	
ii	$1:\frac{1}{2}:4-4\frac{1}{2}$	1:3–4
iii	1:1:5–6	1:5–6
iv	1:2:8–9	1:7–8
v	1:3:10–12	1:8

ability to accommodate movement due to temperature or moisture changes. Cement and sand mixes are generally more resistant to frost attack during construction than cement/lime/sand mixes, but cement/lime/sand mixes give improved adhesion and are generally more resistant to rain penetration. Masonry cement comprises ordinary Portland cement with the addition of a workability aid or plasticiser, so that they can be used as direct alternatives to cement and plasticiser; masonry cement/sand at $1:2\frac{1}{2}-3\frac{1}{2}$ gives a typical type ii mortar, whilst $1:6\frac{1}{2}-7$ gives a typical type v mortar. Workability aids or plasticisers function by allowing air to become entrained in the mixes, improving workability but also improving flexibility and durability.

Although mixes for jointing mortars, renders or screeds are generally specified only in terms of parts of cement and sand, as well as perhaps lime, by volume, these are not the only components that are used in a mix, as water is obviously essential. It is normal to consider that water will be added to a sufficient extent to obtain the desired workability and that this will automatically control the amount of water that is used, but whilst this may be a reasonable assumption in the case of jointing mortars and renders which cannot easily be used unless the amount of water is consistently controlled, screeds can vary much more widely. If the water to cement ratio is excessive, the mortar will suffer high shrinkage on drying which will, of course, be most apparent on render or screed panels rather than on joints. However, if the water to cement ratio is insufficient, the cement reactions cannot occur and the mortar will be weak with inadequate bonding to the background. In fact, water to cement ratios do not depend only on the mix design and the addition of sufficient water to achieve adequate workability; if the background is excessively porous, water and some of the cement components will be extracted from the mortar, causing weakening particularly at the mortar surface in contact with the background and thus severely affecting bonding. This 'suction' problem is best avoided by wetting out the wall background before applying render, or the floor slab before applying a screed, but bricks should also be wetted out by dipping briefly in a bucket of water before laying for the same reason. Unfortunately the necessity for wetting out backgrounds is not generally appreciated and mortar, render and screed bonds are often relatively weak as a result.

Sand for mortar, render or screed mixes is sometimes specified as 'clean and sharp', the traditional way to describe a relatively coarse sand with angular rather than rounded edges, free from clay and very fine aggregate particles. In fact, most of the sand used is ungraded pit sand, with sand from each pit perhaps being recognised as having particular workability properties, but such definitions are insufficiently precise and standard specifications for building sands have been in use for more than 40 years. There were originally three separate British Standard specifications for building sands from natural sources, BS 1198, 1199 and 1200. The current standard was published under these three numbers in 1976 but there have since been a number of major amendments. In 1984 BS 1198 was withdrawn and the title was amended to BS 1199 and 1200:1976 *Specifications for building sands from natural sources*, with BS 1199 covering sands for external renders and internal plasters using lime and cement, and BS 1200 covering sands for jointing mortars. In 1986 an amendment to BS 1199 replaced the original single grading requirement with two grades, type A (coarse) or type B (fine); the particle size limits for types A and B overlap, this overlap representing approximately the earlier limits so that a sand satisfying the earlier requirement should meet both the new type A and B requirements. However, type A sands may be coarser and type B sands may be finer than sand meeting the earlier requirements, and the amended specification stated that:

Experience currently available suggests that satisfactory renderings can be achieved using either grade. Where there is a choice, however, the use of the coarser grade is preferred because finer sands require a higher water to cement ratio which can lead to greater shrinkage than if a coarser grading is used.

The amended combined standard BS 1199 and 1200 was confirmed in 1996.

Where finer sands are used, more water is necessary to maintain workability; this will inevitably lead to higher shrinkage and perhaps loss of bond to the background. Shrinkages occur usually on the outer face of a render, resulting in random surface fractures, but if the bond to the background is weak the panel will become concave with bond failure developing mainly around the edges of each panel. With non-porous backgrounds, such as dense concrete and granite walls, loss of bond and crazing are both likely to develop. Higher shrinkage in renders is inevitable if finer aggregates are used, as additional water must be used to maintain adequate workability, so that control of the water content by measurement is unrealistic, and control is only possible by ensuring that coarse type A sands are used in render mixes.

Precisely the same precautions are necessary with screeds. British Standard BS 8204:Part 1:1987 *In-situ floorings, Part 1, Code of practice for concrete bases and screeds to receive in-situ floorings* suggests that screeds should normally comprise 1:4 cement to fine aggregate, although thicker screeds should comprise fine concrete with 1:4.5 cement to aggregate, with part of the aggregate comprising coarse aggregate up to 10 mm. Fine aggregate is required to comply with grading limit C or M of Table 5 of BS 882:1983, *but with not more than 10% passing sieve size 150 μm*, an added precaution to ensure that the aggregate is not excessively fine, so that excessive water will not be required to maintain workability and the screed will not suffer excessive shrinkage. Floor screed problems are discussed in more detail in section 14.2.

The problems of bonding to porous backgrounds have already been described; the excessive suction removes cementing material from the surface of the mortar in contact with the background, weakening the mortar and the bond. Some migration of cementing materials into the background is essential to establish a bond, the failure being due in this case to excessive migration. However, bonding problems also arise with non-porous backgrounds on which it is impossible to establish a mechanical key in this way. Cement will bond to most non-porous surfaces, such as dense brick, concrete or granite, and it is usually recommended that a splatterdash of neat wet cement should be applied first to improve bonding of rendering. In fact, it is usually more convenient to apply a creamy slurry of cement using a distemper brush. Modern bonding aids can also be used, but they are all based on resins which are slowly attacked by bacteria and they do not have the permanency of cement bonding systems; indeed, the widely used acrylic bonding aids can lose effectiveness over a period of only a few years. Other bonding problems are discussed in more detail in section 11.3.

An unusual and unexpected feature of fresh renders is the way in which they can react with salt in sea spray. The tricalcium aluminate component in the cement will react with the sodium chloride in the sea spray to form chloroaluminate which may result in weakening of the render, and lime in the render (present even if only cement is used) will react with sodium chloride to form calcium chloride and sodium hydroxide, this caustic soda causing stripping off paint if it washes over windows and doors.

7.5 Cast stone masonry

Cast stone is synthetic stone manufactured from concrete. Whether a product is classified as concrete which should meet British Standard BS 6073:Part 1:1981

Specification for precast concrete masonry units or cast stone to BS 1217:1986 *Specification for cast stone* depends only on appearance, the latter specification defining cast stone as 'Any material manufactured with aggregate and cementitious binder and intended to resemble in appearance, and be used in a similar way to, natural stone'. This is, of course, an accurate definition of concrete manufactured to resemble and replace natural stone, and many products will satisfactorily conform with both specifications. However, in an investigation into water penetration problems into a new building in Bath, it was noted that the external skin of the cavity blockwork walls was constructed with a block designed to resemble Bath stone and which therefore blended well with the adjacent limestone buildings, but the stone was excessively porous and dampness in the building was due to excessive water accumulations in the cavities. The blocks were described as concrete resembling natural stone, and it was claimed by the manufacturers only that they conformed with BS 6073:Part 1:1981 as this standard does not specify maximum water absorption requirements; the blocks were not manufactured as conforming to BS 1217:1986 as they could not meet the surface absorption requirements in this cast stone specification. Thus, although the product was marketed as a concrete block resembling natural stone, it could not meet the BS 1217 cast stone requirements, and the claim that it conformed instead to the BS 6073 concrete masonry requirements was a deliberate and misleading attempt to justify the marketing of an unsuitable product. This confusion must be attributed to the British Standards Institution who should make it clear in BS 6073 that concrete masonry intended to resemble and be used in place of natural stone should also conform with the cast stone specification BS 1217. In fact, the Institution introduces further confusion with a third standard BS 6457:1984 *Specification for reconstructed stone masonry units*, which defines such units as 'manufactured with aggregate and cementitious binder, intended to resemble, and to be used for similar purposes as, natural stone'. This specification is therefore precisely similar to BS 1217, except that the latter specification permits the units to be manufactured as a facing resembling stone but on a different concrete material which may be reinforced.

The water penetration limitations which are a feature of BS 1217 and not the other two specifications are intended to limit the amount of water which may enter walls constructed from the cast stone, but the requirements are not intended to indicate complete resistance to water penetration. For example, a new rotunda was constructed using a mixture of cast stone coping units and rather porous French Savonnieres limestone ashlar masonry. Severe staining of the limestone developed just beneath the copings, and this was followed by unsightly lime discharge. Rainwater absorbed into the coping units was percolating downwards into the limestone where it was causing both darkening in the affected areas and permanent staining due to migration of the iron oxide components in the limestone, the lime staining originating from unreacted lime in the cast stone copings. The copings were excessively porous and did not meet BS 1217 requirements in this respect, apparently because the mix was rather dry and had not been adequately compacted into the moulds during manufacture. In fact, this was a deliberate manufacturing technique intended to achieve the required surface texture. The resulting excessive porosity would not normally present a problem in coping units; normal good practice, such as BS 5390 *Code of Practice for stone masonry*, requires a damp-proof course beneath copings, as moisture and thermal movement is usually sufficient to open the joints and allow water penetration, whether the copings themselves are porous or nonporous. However, the unusually dry mix meant that the cement components had not properly reacted, and lime leaching followed, particularly from coping units which were fully exposed to rainfall. The problems at the rotunda were therefore due to the failure to provide a damp-proof course beneath the copings, but also the manufacture of the cast stone from an excessively dry mix. In fact, the copings did not all suffer from lime bleeding, but the copings that were free from this problem also had a 'creamy' surface texture which indicated that a wetter mix had been used. Obviously the 'creamy' texture was not so attractive and the reason for the use of a dry mix was therefore clearly apparent, but this is not an appropriate or reliable way to achieve a change in texture; sufficient water must be provided to ensure complete reaction of the cement components without the use of excessive water, and this mix requirement and surface texture can only be reliably achieved by carefully grading the aggregate.

In all physical respects cast stone performs in the same way as natural stone, so that a microporous structure is much less durable than a macroporous structure. In cast stone these pore size distribution properties will be decided primarily by the aggregate, both by the grading of the aggregate and, where coarser aggregate is involved, the pore size distribution of the aggregate itself. The chemical properties of cast stone are similar to those of mortar and render, as described in section 7.4, and concrete as described in section 8.6, the problems of any reinforcement reactions being described in section 8.5.

7.6 Biological damage

Although most masonry surfaces will be colonised by various organisms, damage is usually restricted to porous stone and is usually associated with atmospheric pollution. The main pollutant today is sulphur dioxide which dissolves in rainwater to produce sulphurous acid and reacts with the calcium carbonate of limestones to form calcium sulphite. However, sulphite is never found on limestones, but only sulphate which is produced by much more aggressive sulphuric acid. Various chemical explanations have been given for this oxidation from sulphite to sulphate, such as ultra-violet radiation, but it is a fact that sulphate deposits are always associated with the presence of sulphating bacteria, particularly *Thiobacillus* species. In areas where coal fires are still used extensively and in some industrial areas, nitrous oxide pollution also occurs which should give nitrous acid and nitrites on limestones, but only nitrates are actually found, indicating oxidation to the more aggressive nitric acid. In these circumstances nitrating bacteria, particularly *Nitrobacter* species, are always found.

Urban areas appear to be cleaner since the introduction of the Clean Air Act, but this is only because particulate emissions have been reduced from industrial chimneys and there has been a progressive decline in most areas in the use of coal for domestic heating. Unfortunately sulphur dioxide pollution has become steadily worse, partly through the increasing use of less expensive heavy oil for heating large commercial and industrial premises as this fuel has a relatively high sulphur content, but also partly through a feature of the Clean Air Act which limits only emission concentrations rather than amounts; if an operator is emitting excessive sulphur dioxide concentrations these can be easily reduced by injecting air into the flue, but the total emissions of sulphur dioxide remain unchanged.

Sulphate formed from sulphur dioxide in this way is a source of crystallisation damage, described in sections 4.4 and 7.2, but damage is not confined to limestones. On mortar the sulphate in urban areas may be sufficient to react with the tricalcium aluminate in ordinary Portland cement to cause the expansion and cohesion failure usually known as sulphate attack. However, deterioration problems attributable to bacteria are not confined to urban areas. In rural areas ammonia generated by bacteria from urine in stables and byres can be absorbed on stone walls or asbestos-cement roofs where it is converted by *Nitrosomonas* species to nitrites and then by *Nitrobacter* species to nitrates, frequently causing spalling damage.

Bacteria are not the only organisms which colonise damp masonry surfaces. If the surfaces are warm with sufficient light, algae will develop in the water film on the surface, typically producing a bright green coloration, although sometimes dark green, brown and pink colorations occur. Algae often colonise a surface within one or two hours of rainfall, but the algal coloration disappears just as rapidly as the surface dries. Many of the algae are killed by drying but sufficient remain to redevelop and multiply when dampness returns. The humus accumulating on the masonry surface from dead algae and other sources eventually allows mosses, liverworts, grasses and even trees to develop, their root systems often causing serious damage. Organic deposits on the surface also encourage fungi to develop, such as *Cladosporium*, *Phoma*, *Alternaria* and *Aureobasidium* species; some species are associated particularly with the high nitrogen levels that develop on masonry contaminated by bird droppings.

Serious masonry deterioration is sometimes associated with growth of lichenised fungi or lichens, symbionts of algae growing within fungi, usually Ascomycetes. The fungal hyphae penetrate deeply into stone, exploring fractures but also generating organic acids such as oxalic acid. Oxalates are formed in carbonaceous stones which are usually deposited in or near the thallus or surface growth; eventually these accumulations of phosphates can kill the thallus, leaving a lichen 'fossil' of calcium oxalate on the surface of the stone which is sometimes mistaken for lichen growth; repeated applications of biocide sometimes fail to control lichen growth because the growth is, in fact, a dead calcium oxalate fossil formed in this way. If the calcium oxalate is deposited just below the surface, densification can occur which is similar in texture to the calcium sulphate densification that can occur on limestones in urban atmospheres, as described in section 7.2, causing similar spalling damage to the surface of the stone, particularly if it is also microporous and subject to frost or salt crystallisation damage. Where lichens grow on roofs, the oxalic and other lichen acids can cause severe damage to lead, copper, zinc and aluminium roof coverings and gutters; these acids can even cause etching on glass and apparently resistant stones, such as granites.

There are basically three types of lichen, classified according to the shape of the thallus. In the crustose lichens the thallus forms a flat crust on the surface of the stone, the diameter of the thallus giving an accurate indication of the age of the growth; the diameter of the largest growths in millimetres will indicate approximately the years since the stone was installed, a useful feature for identifying original and replacement stones in old masonry. The crustose lichens cause densification of the stone surface on limestones and sandstones, the stone within the centre of the thallus often spalling away to leave bare stone which is then rapidly colonised by the growth. Sensitive species cannot develop in polluted atmospheres, but resistant species become very active in the absence of competition, particularly on limestone, cast stone and concrete surfaces on which acid pollutants are neutralised; *Lecanora* and *Candelariella* species are particularly common in these circumstances. Crustose lichens vary greatly in size from minute growths within pores to enormous plates 300 mm (12") or more across.

Foliose lichens have thalli like leaves or scales projecting as a group from a point of attachment to the stone. Fruticose lichens also originate from a point in this way but their thalli are branched. Foliose and fruticose lichens are not so common on buildings, except in exposed and relatively unpolluted areas on western coasts, conditions that actually encourage the development of many different lichens. Particular species tend to be associated with particular conditions. *Lecanora* and *Candelariella* have previously been mentioned as species which tolerate pollution, particularly when growing on acid neutralising substrates such as limestone, carbonaceous sandstone, asbestos-cement tiles, render and concrete. *Calaplaca* species are also commonly found on limestones in reasonably unpolluted conditions, whilst *Tecidia* and *Rhizocarpa* species are more often found on sandstones.

It will be appreciated from these comments that identification of lichen growths can often indicate both the nature of the substrate on which it is developing and the pollution to which it is subject. Very heavy lichen growth on limestone headstones in a cemetery was found to be causing continuous and rapid stone erosion and spalling damage, each sequence of spalling removing the lichen thallus layer with a thin layer of attached stone, thus exposing a fresh stone surface with lettering still engraved in it but with the detail becoming blurred. Identification of the lichen suggest that it was a species which particularly favoured surfaces with a high nitrogen content and which was usually associated with contamination through bird droppings, although none were present on the headstones and all surfaces were virtually identically affected. The explanation for the abnormal growth was pollution through dust discharges from a neighbouring fertiliser factory.

This abnormally heavy lichen growth was associated with Portland limestone headstones, but adjacent memorials constructed in French Euville limestone with a rather different texture developed instead a heavy coating of black slime fungus. Slime fungi are strange organisms which form a heavy gelatinous coating over the stone in which algae are trapped, giving the coating a colour characteristic of the algae involved. Slime fungi can develop externally or internally on building materials. Green, brown or red slime fungi commonly develop on masonry surfaces in churches in humid areas where the periodic heating results in excessive condensation; this is a common problem in churches in Cornwall.

7.7 Cleaning and treatment

If stone or brick and jointing mortar are carefully selected, masonry will be extremely durable and will require little maintenance for structural reasons; even cast stone, concrete and render can be extremely durable. However, where problems are encountered, treatments are available which can alleviate, permanently or temporarily, many of them.

Dirt on limestone surfaces is usually trapped by calcium sulphate deposits, formed as previously described in sections 4.4 and 7.2, and cleaning traditionally involves soaking the surface with water to soften the deposit, followed by cleaning with bristle brushes. The soaking with water must be prolonged, often saturating the structure and perhaps inducing the development of Dry rot during subsequent drying, as described in section 6.4. This cleaning process is very labour intensive and consequently very expensive. In recent years it has been shown that the same cleaning can be achieved more rapidly and with less danger of saturating the structure using high-pressure water jets, although with some harder limestones it is necessary to add a polymer cutting aid to the water. The same technique is often effective on calcareous sandstones, that is sandstones with a calcium carbonate cementing matrix, although the choice of water pressure and cutting aid additive are both very critical with sandstones, as excessive cutting can cause very severe damage. These high-pressure water cleaning systems are also sometimes effective on sandstones with silica cementing, and they should always be tried first when assessing the suitability of a sandstone building for cleaning; severe blackening of sandstone can occur in urban atmospheres and high-pressure water jet represents the least damaging of the cleaning methods that are available. Unfortunately water jets do not always work and it may be necessary to consider alternative systems.

In recent years proprietary hydrofluoric acid systems have been extensively used for cleaning sandstone, apparently because they are the only masonry cleaning system that is currently available with an Agrément Board Certificate. It is emphasised in the Certificate that these products should only be used on sandstones, but unfortunately such cautions are not sufficient; whilst hydrofluoric acid cleaners are suitable for use on sandstones with siliceous cementing, they can cause very severe damage indeed on sandstones with carbonaceous cementing. Products of this type were used for cleaning the tower and spire of a prominent Glasgow church, but they caused severe erosion of the stone surfaces, the enormous quantities of sand generated in this way blocking the drains! Great care must be taken in the use of hydrofluoric acid cleaners, not only because of the dangers to the operatives but also because of the severe damage that can be caused if they are used on unsuitable stone, unfortunately a frequent problem because of the misguided use of Agrément Board Certificates. A much better technique is high-pressure water spray containing a cutting grit; air blasting with sand and husk have been used in the past but a wet process is preferable as it avoids dust problems. In fact, a suitably adaptable high-pressure water jet system can achieve almost all masonry cleaning requirements, although it is essential to adjust pressures and cutting additives to suit particular conditions. Some proprietary systems, such as Jos with its swirling action, can be particularly effective.

It is not generally appreciated that much of the dirt on buildings is, in fact, biological growth; even black 'soot' is often a mixture of lichen and fungal growth. It is therefore always sensible to carry out trials with suitable biocides before deciding to use a cutting method; even high-pressure jets with plain water cause erosion of a masonry surface. In addition, one of the problems with wet cutting techniques is the rapid development of algal growth following cleaning, so that a biocide application should automatically follow any wet cleaning technique. Traditional biocides include chlorine, introduced as sodium hypochlorite solution or calcium hypochlorite (bleaching powder) paste, various heavy metals such as copper and zinc, and phenols. Hypochlorite is only a sterilant without any prolonged inhibitory action. Metal salts are also very transient in their action, and copper salts generally produce green or blue staining which is unacceptable on light coloured masonry. Phenols are both sterilants and inhibitors, although the inhibitory action is far from permanent. In recent years chlorophenols, particularly sodium pentachlorophenate, have been very popular, although they are not particularly effective and their use can cause serious problems. They are reasonably efficient as sterilants but they decompose on masonry surfaces and also react with any iron that may be present, causing brown or mauve staining.

About 30 years ago the author's laboratory was commissioned by the Commonwealth War Graves' Commission and subsequently by the Directorate of Ancient Monuments and Historic Buildings (now English Heritage) to investigate systematically biocides for masonry. It was concluded as a result of this work that there were three groups of biocides which were particularly effective. Borates, most suitably applied as the proprietary mixture of borax and boric acid known as Polybor to obtain optimum solubility, are easy and safe to apply to masonry with no risk of any staining damage. Following application, the sodium ion is neutralised by carbon dioxide in the air, leaving a deposit of boric acid with only limited water solubility; this has proved to give prolonged protection against algal and lichen growth. Quaternary ammonium compounds have also proved to be safe, effective and reliable, the most suitable compounds being the groups that are readily soluble in water, particularly the benzyl alkyl dimethyl ammonium compounds available as benzalkonium chloride, and alkylbenzyl trimethyl ammonium compounds which are not now readily available. These compounds are reasonably persistent on masonry, but greater permanence can be achieved by adding a cationic biocide such as tri-n-butyltin oxide (TBTO); this compound is an oil which is immiscible with water and therefore unsuitable for this application when used alone, but it can be solubilised by mixing with a quaternary ammonium compound, the combination broadening the spectrum of activity and producing a biocide which is much more effective than either of the two component compounds. Unfortunately, the use of

TBTO is now restricted but it can be replaced by other cationic biocides such as some zinc compounds.

Water repellents can be used to control biological growth by reducing rainwater penetration into masonry; their use has already been fully described in section 4.11.

There have been many attempts over the years to consolidate friable masonry. One of the oldest systems involved the treatment of friable limestone surfaces with lime water, the lime carbonating and thus adding calcium carbonate to the limestone. The system was unpredictable and usually ineffective, as only weak solution concentrations can be obtained and even repeated applications achieve only limited consolidation, mainly concentrated close to the treated surface. In fact, one of the disadvantages with such processes is the way in which they densify the treated surface, the subsequent spalling of the densified surface often causing more severe damage than the progressive erosion that the consolidant was designed to prevent. Baryta water, in which the calcium hydroxide in lime water is replaced with barium hydroxide, is much more reliable for two particular reasons. Barium hydroxide is much more soluble than calcium hydroxide and it is much easier to achieve reasonable solution concentrations. In addition, barium salts formed in polluted atmospheres are much less troublesome than calcium salts. For these reasons baryta water should always be used in preference to lime water as an inorganic consolidant system.

Various resins and waxes have been proposed as consolidants over the years, but even modern resins such as polyesters, polyurethanes and epoxides must never be used as they are totally unsuitable. If they are applied at sufficient loadings to achieve consolidation they invariably seal the surface and induce serious spalling problems. In addition their durability is poor; whilst modern resin systems are often very durable when applied in reasonable bulk, they deteriorate rapidly when applied as dispersed treatments on masonry surfaces exposed to intense sunlight and weathering.

Only silicon chemistry can produce systems which avoid these various problems. Silica deposits can be formed in friable masonry surfaces in various ways and can prove extremely durable. The most suitable systems are based on tetra-alkoxy silanes, sometimes known as siliconesters or alkyl silicates; these systems react with water, releasing alcohol and depositing a silica polymer which cements the friable masonry components in the same way as a silica cementing matrix in a sandstone. Obviously such consolidants tend to densify the treated surface, and may result in surface spalling, so that it is always advisable to follow consolidation treatment with a water repellent treatment to minimise water penetration and to reduce the risk of spalling; combined consolidant and water repellent treatments have been developed, such as the Brethane treatment developed by the Building Research Establishment following work sponsored by the Establishment in the author's laboratories. The treatment involves the application of a monomer to the masonry, which then polymerises in situ to provide the consolidant action, but the system is not easy to apply as the monomer is very volatile. Partly polymerised systems can be produced which are likely to be more effective, although they are not commercially available at the time of writing.

8

Concrete problems

8.1 Introduction

Concrete technology is much too complex to be discussed in detail in a short chapter of this book. Instead the following account will be confined to a discussion of the various problems that may be encountered in concrete used in buildings.

The first use for concrete in buildings was in the construction of ground floor slabs, originally unreinforced but now almost always steel reinforced, as it enables a thinner slab and less concrete to be used to achieve the same strength and reliability. Reinforced concrete was subsequently used for the construction of suspended floors, and eventually for beams and posts. In many countries poured concrete is also used for the construction of walls, but this is unusual in the British Isles where walls are usually constructed in masonry, that is blocks of material joined by mortar, although in major buildings the walls often comprise curtain walls or filling panels in association with a reinforced concrete structure. Concrete blocks are not considered in detail in this chapter but instead in section 7.3 of the previous chapter on masonry.

This chapter does not consider failures, defects and other problems arising through faulty design of concrete structures in terms of dimensions, mix or reinforcement content, as these are fundamental design factors rather than problems of a type that require special investigation.

8.2 Mix design

Concrete mixes were originally specified in volume proportions as concrete was prepared in this way, but batching by weight was found to give much more consistent mixes with more reliable results; concrete is now prepared only in this way, although mortar for jointing and rendering is still usually prepared on site by volume. In recent years it has become less usual for engineers to specify an actual mix, but instead to specify a particular strength designation, the contractor being required to demonstrate using test blocks that their chosen mix will achieve the strength requirements.

These progressive changes in specification method have simplified the situation for design engineers but complicated the situation for contractors. Several problems have developed which can be attributed to this change of emphasis in specification which encourages contractors to adopt a policy of 'if in doubt add more cement' to ensure that the concrete meets the specified strength requirements, the various guidance documents emphasising the importance of an adequate cement content. In fact, excessive cement may lead to high shrinkage; this will either manifest itself as shrinkage of the concrete elements during curing, or as fractures or crazing of the surface of the concrete. The water to cement ratio is also critical, and if the amount of water is excessive it can encourage excessive shrinkage.

Concrete is normally prepared using ordinary Portland cement (OPC), although for many purposes a proportion of the cement can be replaced without detriment by other materials, such as pulverised fuel ash or ground granulated blast furnace slag. Sulphate-resisting Portland cement is a direct alternative to ordinary Portland cement which is used when there is a danger of sulphate attack; it differs mainly by having a much lower tricalcium aluminate content, the component that reacts with sulphate. High alumina cement was originally developed by a Frenchman Jules Bied as an alternative to ordinary Portland cement where resistance to sulphate attack was required, although this product, which is also known as Ciment Fondu or Lightning cement, was mainly used where development of high early strength was required. For this reason high alumina cement was widely adopted for the manufacture of precast concrete units, including some system buildings but particularly proprietary floor and roof units comprising beams and slabs or beams and infill pots. However, in warm humid conditions chemical conversion of the cement to alternative forms can occur, resulting in substantial loss of strength. This loss of strength has resulted in some spectacular structural collapses, firstly in France commencing in about 1930; however, as these collapses were not directly linked with the use of high alumina cement, the product continued in use and was being utilised to an increasing extent in the British Isles when a series of collapses occurred which prompted doubts about the reliability of high alumina cement concrete. The main failures involved the Assembly Hall at the Camden School for Girls, the Bennett Building at the University of Leicester, and the Sir John Cass's Foundation and Red Coat Church of England Secondary School in Stepney. It is true to say that these and other collapses occurred because the concrete roof structures were excessively stressed, so that loss of strength in the concrete resulted in collapse, and collapse would have been much less likely if the design requirements had been less critical and incorporated normal safety factors. In particular, beams had very limited bearings at their ends and it was the extreme stressing of these bearings that resulted in the collapses.

These spectacular accidents resulted in fears that any concrete component based on high alumina cement could similarly collapse. Components in buildings were extensively analysed and replaced where high alumina cement was found, although many replaced components such as lintels were far from over-stressed and there was no real justification for their replacement.

8.3 Carbonation

Ordinary Portland cement is extremely alkaline, mainly because of its normal lime content but also because some cements contain appreciable concentrations of sodium and potassium salts. These alkalis are progressively neutralised by the absorption of carbon dioxide from the atmosphere, the reaction occurring more rapidly in damp concrete. This carbonation and the loss of alkalinity is mainly important in relation to the durability of steel reinforcement, as it is the alkalinity that prevents steel corrosion in concrete.

The rate of carbonation is therefore very important in relation to the durability of reinforced concrete; it depends on the rate of diffusion of carbon dioxide into the concrete, as well as the amount of moisture that may be present in the concrete. Generally, dense cement-rich concrete carbonates much more slowly than porous weaker concrete because density limits both carbon dioxide diffusion and rainwater absorption, although it is essential to appreciate that higher density only slows the rate of carbonation and does not prevent it; reinforced concrete can never have an indefinite life. As a general rule sufficient life is obtained by specifying a minimum concrete cover over reinforcement to delay reinforcement rusting for a sufficient period; reinforcement is discussed in more detail in section 8.5. Depth of carbonation is easily checked on the broken surface of a sample of concrete or in a hole drilled in solid concrete, by applying a light spray of phenolphthalein reagent which gives a bright red colour on alkaline cement but is colourless on carbonated cement.

8.4 Aggregate

Concrete consists of aggregate which is bonded together to form a solid material, and the proportion, grading and type of aggregate have a profound influence on the properties of the final concrete product. Aggregate grading is particularly important. Concrete is generally strongest, most stable and least expensive if it uses the largest possible aggregate, but the largest aggregate size must not be more than two-thirds of the smallest space that the concrete is required to fill, the critical dimensions usually being the cover over reinforcement, the minimum space between reinforcement components or the depth of a slab. Ideally aggregate should be continuously graded so that it contains an appropriate proportion of smaller particles to efficiently fill the gaps between larger particles, although in practical terms it is usual to mix large and small aggregate in appropriate proportions to reasonably satisfy this requirement. Small or fine aggregate is itself available in a series of grades which can be further mixed to improve overall aggregate grading, although it is important to avoid the use of excessively fine aggregate, or aggregate containing dust or clay particles. If excessively fine aggregate is present it will be necessary to increase the water content to maintain workability, and the resulting increase in the water to cement ratio will then result in excessive shrinkage as described in section 8.2.

Shrinkage in concrete on drying is generally about 0.03% plus any additional shrinkage which may be attributed to the use of excessive water or the use of a shrinkable aggregate, that is an aggregate which expands on wetting and shrinks on drying. Stable aggregates such as quartz sand, flint gravel, marble chips and dense granite chips can actually produce concrete having a shrinkage of less than 0.025%. However, limestone aggregates vary from small to medium shrinkage, and sandstones from medium to high shrinkage, whilst some other sedimentary rocks such as greywacke, shale and mudstone may result in concrete shrinkage exceeding 0.085%. Aggregates from some igneous rocks of the basalt and dolorite types can also suffer high shrinkage, although some dolorites are virtually non-shrinkable so that the situation is very confusing, particularly in parts of Scotland where mixtures of shrinkable and non-shrinkable aggregates often occur unpredictably in glacial deposits which are utilised as aggregate.

Concrete with movement up to about 0.065% is suitable for most purposes, although care should be taken in the design of thin units, such as cladding panels and cast floors with movements in excess of 0.045%. Concrete with movement in excess of 0.065% is suitable for general structural purposes, although not for unreinforced or thin reinforced structures; severe deterioration can occur when such concrete is exposed to the weather, although durability can be improved by air entrainment. Concrete with movement exceeding 0.085% is suitable only for use where drying out never occurs, or where the concrete is heavily reinforced and not exposed to the weather. It will be appreciated from these comments that, whilst initial shrinkage may be very significant in some circumstances, particularly in relation to differential shrinkage in structures where, for example, the concrete shrinks to a much greater extent than cladding or other components, the most serious damage is caused by alternating shrinkage and expansion through wetting and drying cycles, surface damage being most apparent but loss of strength also occurring. It is therefore recommended that all aggregates should be assessed for shrinkage before they are used. Shrinkage tests are advisable whenever a new aggregate is used, and are essential whenever there are any changes in mix design or sources of aggregate in areas in which shrinkable aggregates are known to occur.

The ultimate surface of a concrete component is usually cement-rich, largely concealing the aggregate, a surface that is achieved automatically if concrete is well compacted against the mould or the upper free surface is worked in any way. However, some concrete, particularly decorative panels or wall units, is designed to have an exposed aggregate finish, achieved by washing the cement paste away when the concrete is semi-dry using water jets or sometimes by using weak acid, although acid cannot be used with limestone or carbonaceous sandstone aggregates. The choice of aggregate is critical. It must have the same durability as would be necessary for normal masonry, but it must also adhere durably to the supporting cement paste, and it must be virtually non-shrinkable to avoid danger of any detachment from the background concrete. Exposed aggregate has a very pleasing texture but it is difficult to achieve consistent performance for these various reasons. A more realistic approach is to cut back the surface with abrasive disc or belt grinders to expose an aggregate surface that can be polished, although the exposure of the aggregate again makes it necessary for it to be selected with considerable care.

8.5 Reinforcement

This section is not concerned with any defects or failures due to reinforcement design, but only with problems that generally affect reinforcement in concrete. Although a variety of reinforcements are available, including mineral, glass and carbon fibres, as well as various metal meshes and rods, steel is most widely used and the most serious problems are associated with steel corrosion.

Steel corrosion can only occur in the presence of moisture and has already been described in detail in section 4.5. The high alkalinity of the cement paste surrounding the steel reinforcement normally inhibits corrosion, but the alkalinity is progressively lost by carbonation as described in section 8.3 and steel corrosion can eventually develop. A reasonable period of freedom from corrosion is normally achieved by specifying a minimum concrete cover over the reinforcement, usually 50 mm (2"), although the life that this achieves will depend upon the density of the concrete and the porosity of the aggregate.

If cover is inadequate over reinforcement, or even over tie wires, corrosion becomes apparent as unsightly patches of brown staining and brown water runs, but progressive corrosion of reinforcement rods results in expansion and spalling of the concrete surface which is unsightly, as well as being destructive and dangerous. Various methods have been proposed for inhibiting this damage; they usually involve cutting out the spalling concrete to expose fully the corroded reinforcement, removal of the corrosion and treatment of the reinforcement with tar, bitumen or resin coatings in an attempt to prevent further corrosion, followed by patching the spalled concrete, usually with resin repair systems. Repairs of this type are not recommended for a variety of reasons. Resin repairs to concrete are, of course, incompatible and always weather differently from the original concrete, resulting in an unsightly patchy appearance that can only be avoided by completely decorating the concrete. Although coal tar systems give good corrosion inhibition, they are only effective on the reinforcement to which they are applied and there is a danger that spalling will develop later adjacent to the repaired area. Bitumen and resin systems are much less efficient in inhibiting corrosion. The most reliable system is, in fact, to revert to the original design concept by exposing the corroded reinforcement and cleaning off the corrosion as far as possible, followed by application of a creamy cement slurry to the reinforcement and spalled surfaces, and filling the spall with a matching concrete mix, a repair that will weather to match closely the original concrete and which will restore the alkalinity and corrosion resistance of the original system.

These comments are, of course, applicable to the normal oxidative corrosion of steel which is usually described as rusting, but other forms of corrosion can also occur. If different reinforcement metals are mixed, electrochemical corrosion is probable, but such corrosion can also occur if a single piece of reinforcement is exposed to an electrolyte that varies in nature, a situation that can frequently occur in large structures, such as bridges, where the general dampness in the concrete represents the electrolyte in which abnormal concentrations may occur, for example, through local concentrations of de-icing salts. Such problems are not common in buildings, but they are mentioned here as they will be difficult to diagnose if they occur.

8.6 Chemical damage

Carbonation due to absorption of carbon dioxide from the atmosphere is the most important and inevitable form of chemical damage that affects concrete. Carbonation and its effect on steel reinforcement corrosion have already been described in sections 8.3 and 8.5, but there are several other forms of chemical damage which are frequently encountered in concrete.

Calcium chloride was extensively used in the past in concrete as a cement setting accelerator. It acts by increasing the calcium ion concentration of the mix, but it also introduces chloride ions which can accelerate steel corrosion. It was considered until recently that the presence of chloride was only detrimental in this way when corrosion was caused by some other factor, such as inadequate cover or deep carbonation resulting from excessive permeability of the concrete, but it is now known that reinforcement corrosion is likely to be caused by chloride, even at relatively low moisture contents and even if the surrounding concrete remains alkaline. Building Research Digest 264 suggests that the risk of reinforcement corrosion is low if the chloride ion content on the cement is less than 0.4%, or medium for 0.4–1.0%, or high above 1.0%. These guidance figures are based on the hypothesis that, when calcium chloride is added to concrete as an accelerator, some of the troublesome chloride will combine with the cement as a complex which is stable under alkaline conditions. The amount of chloride that can be immobilised in this way, and which will only contribute to corrosion in the event of carbonation reaching the reinforcement, is a function of the composition of the cement, particularly the tricalcium aluminate content, so that the risk of corrosion will be lower with cements which have relatively high concentrations of this component. Sulphate-resisting cement has a particularly low tricalcium aluminate content, and the risk of corrosion in the presence of a calcium chloride accelerator is therefore enhanced. The important point is that there is always a risk of steel reinforcement corrosion where concrete has carbonated to the depth of the reinforcement, and this corrosion will be more severe if chloride is also present; but there is also a risk of corrosion even if carbonation has not occurred if excessive chloride is present, particularly if ordinary Portland cement has not been used and has been replaced by sulphate-resisting cement or even a mixture of ordinary Portland cement with pulverised fuel ash, these changes reducing the tricalcium aluminate content and reducing the ability of the cement to immobilise the troublesome chloride.

Chloride can also be introduced into concrete in sea spray, causing precisely the same risk of steel reinforcement damage. Sea water contains high concentrations of sodium chloride. If concrete, render or mortar is fresh and uncarbonated the chloride ions will react with the tricalcium aluminate in the cement to form calcium chloroaluminate, the free sodium ions released by this reaction then increasing the alkalinity of the concrete. Washings from the concrete will therefore contain sodium hydroxide (caustic soda) which can severely damage paint coatings and various metals. This high alkalinity can also cause problems through reaction with the concrete aggregate. Alkali silica reaction is most common and involves siliceous aggregates, such as sandstones, which react to form a calcium alkali silicate gel which absorbs water and expands, disrupting the concrete. A similar alkali carbonate reaction can occur with certain argillaceous dolomitic limestones, although it is not common in the British Isles. Whilst alkali may arise through the reaction of sea water with uncarbonated cement as previously described, some cements also have abnormally high sodium or potassium contents which can cause similar problems when they are used in conjunction with susceptible aggregates. Alkali reactions cause expansion and fractures, either random fractures in large panels or longitudinal fractures in posts and beams; the latter fractures are often similar to those caused by reinforcement corrosion, although exposure of the reinforcement is usually sufficient to demonstrate that corrosion is not significant and not the cause of the damage.

Alkali silica reactions caused by the use of high alkali cement in conjunction with susceptible aggregate can cause dramatic concrete deterioration, sometimes described as 'concrete cancer'. Problems have been encountered, particularly in Devon and Cornwall, where locally produced susceptible aggregates have been used in conjunction with locally produced high alkali cement, although problems are not restricted to this area; some of the aggregates used in Jersey have also proved to be susceptible to this reaction.

Concrete can deteriorate through sulphate attack if it has a sufficient moisture content. Sulphate attack has been described in detail in section 4.4 and affects particularly brickwork and render where the sulphate is derived from bricks. Concrete structures may be affected by sulphates in the soil, particularly in industrial areas where ash and clinker wastes often contain high sulphates. In fact, the most common cause of sulphate attack in concrete is on ground floor slabs laid over ash or clinker fill without the protection of an isolating damp-proof membrane, so that sulphate solution can be slowly absorbed into the concrete slab. Initially the sulphate reacts with the tricalcium aluminate in the cement to form calcium sulphoaluminate which absorbs water of crystallisation to cause substantial expansion, sometimes as much as 10%. The expansion of the lower side of a floor slab in this way eventually causes cupping distortion and fractures. However, the sulphoaluminate is only stable in alkaline conditions, decomposing to calcium sulphate as the concrete eventually carbonates, destroying the cohesiveness of the concrete. In a floor slab, sulphate attack and subsequent carbonation occur very slowly and damage may not become apparent for many years.

A different form of sulphate attack can occur in concrete and brickwork mortar in sewers. Generally, the sewerage has a high organic content and is anaerobic. Sewerage normally contains sulphates, which are reduced by sulphate reducing bacteria, such as *Desulfovibria* species, to hydrogen sulphide, commonly known as 'bad egg' gas. The hydrogen sulphide is released into the air space above the sewerage which is aerobic and, as the hydrogen suphide is absorbed into water films, it is oxidised by sulphating bacteria, such as *Thiobacillus* species, to form ultimately sulphuric acid which reacts with the tricalcium aluminate in the cement in exposed concrete or mortar to form calcium sulphoaluminate as previously described:

$$3CaO.Al_2O_3 + 3H_2SO_4 + 3Ca(OH)_2 + 25H_2O = 3CaO.Al_2O_3.3CaSO_4.31H_2O$$

The considerable expansion caused by this reaction results from the absorption of 31 molecules of water of crystallisation on the calcium sulphoaluminate, and in sewers this expansion followed by carbonation and weakening results in progressive spalling erosion of concrete surfaces.

9

Coating problems

9.1 Introduction

Wood, metal and masonry coatings are normally required to perform a decorative or protective function. Coating technology is extremely complex and will not be considered in detail, this chapter being concerned only with coating defects which may be observed in the normal course of building defect investigations. This chapter is therefore only a brief review drawing attention to information available in other parts of this book.

9.2 Wood coatings

Wood varnish finishes are both decorative and protective in function. They achieve their decorative effect optically through their refractive index, coated wood appearing to be, in effect, embedded in a glass-like material. This has the effect of changing the appearance of the wood surface, darkening the apparent colour but also making the features of the wood more readily visible.

Varnish can give a very attractive and virtually permanent finish in interior situations, such as on furniture out of direct sunlight, but exposure to ultra-violet light generally causes oxidation and the coating becomes brittle and eventually opaque. Some modern varnishes have greatly improved resistance to this form of deterioration, but almost all varnishes suffer from preferential wetting failure; wood is hydrophilic whereas varnishes are hydrophobic, and if water accumulates at their interface it will tend to spread over the wood elements, dislodging the varnish. The varnish coating then appears to become opaque and unsightly as it separates from the wood elements, although this is actually an optical effect caused by reflection at the rear surface of the coating following separation from the wood.

The conventional way to reduce preferential wetting failure is to establish a reliable mechanical key to the wood by encouraging penetration. Usually the first coat of varnish is thinned to reduce viscosity but this actually has very little effect on penetration; although the viscosity of the applied solution is reduced the molecular size of the varnish components are unchanged and still tend to be deposited on the surface of the wood with only the diluent solvent penetrating more deeply, and only the use of a varnish primer with lower molecular size components will significantly improve the situation. Some organic solvent wood preservatives designed for application prior to varnish can be reasonably effective in this respect, as they may contain low molecular weight resins which can act in this way, particularly if they are applied by impregnation, such as a double vacuum cycle. However, completely reliable resistance to preferential wetting failure can only be achieved using chemical bonding agents to link the varnish to the wood elements. Obviously the bonding agents themselves must be resistant to deterioration by moisture or sunlight and this severely limits the availability of suitable systems. Wood preservation technology can give guidance in this respect, as it is necessary to find compounds which will fix to the wood elements to provide a modified surface to which the varnish will adhere without being affected by preferential wetting. Some amines, particularly quaternary ammonium compounds, can be very effective when applied as primers, usually at appropriate concentrations in water with thorough drying before the varnish is applied. Organotin compounds can also be very effective, although they are not normally used at sufficient concentration for this purpose in ordinary organic solvent wood preservatives. Some silicone resins can also be particularly effective, provided that they have sufficient functionality; BS 3826 Class B resins are usually suitable. Whilst these systems are best applied in light organic solvent, perhaps with the addition of a limited amount of petroleum resin as a primer, the concept of a separate priming system for varnish is not popular with manufacturers, although it is normally accepted for paint systems. Some manufacturers have introduced integrated preservative, stain and finish systems which are, in effect, primers followed by varnish finishes; they generally perform much more reliably than normal varnish systems and they are also much easier to maintain, usually requiring only solvent cleaning or light sanding and the application of a further finish coat as a maintenance treatment. However, if a deep and glossy finish is required, only conventional varnish systems will be suitable, but there are wide differences in durability and ease of maintenance. Conventional 'spar' or yacht varnishes are not particularly durable, but they remain reasonably soft and they are easy to strip and recoat; yacht maintenance is usually annual, whereas external building maintenance occurs normally at 5 year or longer intervals. Some modern systems, such as two-pack polyurethanes, give remarkable resistance to water penetration; this is not always desirable, however, as humid air diffusing from the interior of a building will result in the accumulation of condensation beneath external coatings which will encourage preferential wetting failure. Polyurethane coatings of this type are also usually very hard and very difficult to remove should the finish become opaque through preferential wetting failure, so that finishes of this type should never normally be used as varnishes for exterior application, although they are widely promoted for this purpose. One-pack moisture cured polyurethanes are much more suitable; they have similar resistance to water penetration, but they are generally softer and with many systems their polymerisation reactions result in the establishment of chemical bonds to the wood substrate and good resistance to preferential wetting failure.

Paints are essentially pigmented varnishes so that the colour of the finish depends on the pigment content rather than the substrate to which they are applied. They suffer from the same failures as varnishes due to oxidation in sunlight and loss of adhesion through preferential wetting, although the pigments generally absorb sunlight and reduce the rate of development of oxidation damage; usually only the varnish binding the exterior pigment is significantly affected, resulting in loss of gloss and 'chalking' of the surface but without severe damage to the coating system if it is sufficiently thick. The better performance of paints in this way has encouraged research on the addition of transparent pigments to varnishes, silica flour being one system that has proved effective.

Paint coating failures are generally due not to failure of the final gloss finish, but to inadequacies in the undercoat or primer. The function of the primer is usually said to be to seal the porosity of the wood surface and conventional primers in recent years have contained high pigment loadings to achieve this function. In fact, the most important function of a primer is to establish adhesion to the wood substrate, and the most reliable primers tend to be based on low molecular weight varnishes containing comparatively low concentrations of pigments as the filling of the surface porosity can be achieved just as effectively by the high pigment content in the subsequently applied undercoat. It is probable that primers with high pigment loadings are preferred because they have a good masking effect, apparently a valuable asset when joinery is primed before delivery to site; masking can be achieved, even in the absence of any pigment in a primer, by the undercoat which is intended to provide a smooth background of uniform colour to which the finish can be applied, so that the undercoat must have a high pigment content coupled with adequate varnish to ensure good cohesion. The finish must adhere reliably to the undercoat and must give the desired appearance, coupled with adequate durability.

Unfortunately ideal painting is seldom achieved. On window frames problems occur at joints, where differential movement between side-grain and end-grain always results in the development of a fracture which allows water to penetrate beneath the coating. At one time, high quality joinery joints were coated before assembly to minimise this effect, but double vacuum or immersion treatment with organic solvent wood preservatives has greatly reduced problems by preventing fungal decay damage through this water penetration; the use of water repellent preservatives also minimises water movement beneath the coatings and extending finish life by delaying or preventing preferential wetting failure, which normally causes blister or peeling damage.

The development of black staining beneath varnishes is usually described as 'stain in service', in contrast to sapstain or bluestain in freshly felled wood, as previously described in section 6.4, although some fungi can cause both types of staining. These staining fungi are generally resistant to the fungicides that are used in preservatives designed to prevent decay by wood destroying Basidiomycetes, and stain problems can only be avoided by including other fungicides in preservatives, primers and finishes. Although these stain fungi develop on the wood substrate, they can also cause deterioration of the varnish, typically forming holes through to the surface which support minute black pustules in the case of the most troublesome species *Aureobasidium pullulans*. It is not generally appreciated that stain problems and particularly this species are also troublesome on paints, causing the same boreholes through the coating and darkening of light surfaces as pustules accumulate. This form of damage can usually be identified using a hand lens, although an illuminated 30× microscope is particularly useful for this purpose.

These comments have been concerned with conventional varnish and paint systems. Many problems can be attributed to a failure to apply coatings of sufficient thickness, or the use of several coats of undercoat but only a single coat of finish, whereas a single coat of undercoat and several coats of finish would give a much more durable result. Rapid loss of adhesion and peeling failure can often be attributed to the use of an unsuitable primer, particularly the inexpensive 'wash' emulsion primers often applied by joinery manufacturers. These water-based systems do not achieve reliable adhesion to the wood, but they are also water sensitive and their expansion and contraction with seasonal changes in moisture content can severely stress subsequently applied paint systems, causing premature failure.

9.3 Metal coatings

The use of paint coatings in metal corrosion prevention has already been described in section 4.5. Any damage to a metal coating exposed to water and air will result in corrosion just beneath the coating around the damaged area. In other respects the reliability of a coating depends on its adhesion to the metal substrate, and great care is necessary in cleaning the substrate before priming and in the choice of primer, as well as in achieving an adequate thickness of finish. Failures can always be attributed to a lack of care in one of these three respects.

The most suitable primers for use on iron and steel are zinc chromate and zinc rich coatings (often described as cold galvanising). Where paint is already damaged through corrosion, removal of rusting, followed by a zinc rich coating to the damaged areas and overall application of undercoats and finish coats, probably represents the most convenient and reliable maintenance technique.

The normal excellent durability of aluminium when exposed to the weather is due to the formation of a protective oxide over the surface, and this protective system is particularly well developed in anodised aluminium. However, these oxide coatings also interfere with paint adhesion and they must be completely removed immediately prior to the application of primer in any painting operation on aluminium. The most common paint on aluminium in buildings is on window and door frames, and it is normally applied to the aluminium extrusions before assembly. Simple immersion is not very efficient as sag and runs occur, as well as thin films on sharp edges. These problems are completely avoided with acrylic coatings applied by electrophoretic deposition, a process that ensures a uniform coating of controlled thickness, even on sharp edges. However, the process is particularly sensitive to contamination of the aluminium surface prior to coating, and acrylic coatings of this type are not usually very durable, suffering surface chalking. In contrast polyester powder coatings, in which the aluminium surface is loaded with dry powder by electrostatic deposition and then baked in an oven to fuse the powder into a continuous coating, are much more durable. A few years ago acrylic electrophoretic coatings were widely used, as they were much less expensive than polyester powder coatings, but the price differential is now much smaller and it is unwise to use acrylic coatings now that much more reliable and more durable polyester coatings are available at reasonable cost.

There is only one other type of metal coating that is extensively used in domestic buildings, and that is the lacquer on brass door furniture and similar fittings. Brass tarnishes rapidly and can only be maintained in bright condition if it is frequently polished. However, an alternative is polishing during manufacture followed by application of a lacquer to prevent tarnishing and preserve the polished appearance. The lacquer must be tough and applied sufficiently heavily to resist abrasion and chipping in normal service, but many commercial lacquers are applied much too thinly and wear away in places giving the fittings an unsightly patchy appearance. For external fittings it is very important that the lacquer protection should completely prevent water penetration beneath the film, but lacquer films are often too thin to be reliable with 'holidays' or perforations due to air bubbles during application. In addition the usual acrylic lacquers that are used have poor resistance to sunlight, eventually crazing and allowing water penetration which occurs first in abraded locations leading to unsightly local staining. This staining is particularly severe in external fittings exposed to even slight sea spray or urban pollution, and can develop particularly rapidly with the very thin acrylic lacquer coatings that are normally applied to brass door furniture and similar fittings. The obvious advice in these circumstances must be to use lacquered brass fittings externally only in situations where they will be protected from rainfall and direct sunlight, and to use lacquered brass generally only in situations where it will encounter light wear; in all situations non-lacquered brass fittings will prove more satisfactory, although they obviously need regular polishing.

9.4 Masonry coatings

Interior coatings normally deteriorate only if they are affected by dampness. In such circumstances the main priority must be to eliminate the dampness, as advised in Chapter 4, and problems with interior coatings will not therefore be discussed in detail in this chapter.

Exterior coatings were originally lime washes, although they are not very durable, particularly in polluted urban conditions. Hydraulic lime, usually manufactured from lias limestone containing clay and therefore comprising a natural cement, give much more durable finishes. However, the most durable finishes were found to be obtained using lime and tallow, usually with sea water or sea salt added to assist emulsification, the salt being hygroscopic and maintaining the moisture content and flexibility of the coating with obvious advantages in durability. These systems were progressively improved and became the oil-bound distempers, originally involving oil emulsions stabilised with milk which also contributed casein binders, but these systems have now been superseded by modern acrylic and similar emulsion paints. One of the problems that affects both interior and exterior plaster, render and masonry finishes is a succession of paints, the older coatings often losing adhesion and the drying shrinkage of the modern emulsion paint often encouraging this loss of adhesion. In such circumstances, it is important to scrape and wire brush the surface to remove all loose paint as far as possible, to apply liberally a primer of emulsion paint thinned with an equal volume of water, followed by several coats of full concentration paint, perhaps with sanding between coats on smooth surfaces to conceal flaking of earlier coats.

Experience with hydraulic lime naturally prompted the use of Portland cement slurries as masonry paints, and eventually the development of more decorative finishes by using white cement and added pigments. The durability of a cement coating of this type depends largely on the adhesion to the substrate, particularly on very porous surfaces which should be wetted immediately prior to painting to kill the suction which can remove much of the soluble cement binder from the paint coating. Cement shrinks on curing and this can stress the coating, causing fine crazing. This problem can be reduced by using a mixture of cement and very fine aggregate, but the addition of suitable polymers can result in a coating which is easier to apply, better bonded to the substrate and more resistant to crazing as it is more flexible. The polymers are usually similar to those used in emulsion paints, but the combination with cement results in a product which is much more satisfactory and very much more durable than an emulsion paint.

Emulsion and cement paints are reasonably permeable and any penetration or condensation moisture accumulating beneath the coating is able to slowly disperse by evaporation. If impermeable coatings are used on masonry, trapped moisture cannot disperse in this way and severe spalling damage can occur during freezing conditions. Such damage was associated in the past with tar or bitumen coatings used in an attempt to reduce rain penetration into walls, although today such damage is usually associated with modern polyester, epoxy or polyurethane finishes which are completely unsuitable for use on masonry. There is, however, one other type of finish which is extensively used on masonry and which can be formulated to give permeable or impermeable coatings. Chlorinated rubber, which is available under various proprietary names, is usually formulated as an emulsion to give permeable masonry finishes, but it is also available as an organic solvent solution which can achieve a very impermeable coating, particularly suitable for use as an underwater decoration for swimming pools.

Vapour permeability of structural surfaces and coatings involves various units:

Vapour resistance	MNs/g (also written sMN/g)
Vapour permeance	g/MNs (also written μg/Ns)
Vapour resistivity	MNs/gm
Vapour permeability	gm/MNs (also written μgm/Ns)

The most convenient starting point is vapour permeance, which concerns the amount of water vapour (g) which can be transmitted depending on the pressure (MN) across the material and the time (s) of transmission. Vapour permeability introduces the area (m²) over the thickness (m), cancelling out to introduce the single unit of length (m). Vapour resistance and resistivity are simply the inverse of permeance and permeability, so that resistance and permeance are reciprocals, as are resistivity and permeability, but only if these standard units are used. Moisture vapour permeability is sometimes described as moisture vapour transmission and expressed as $g/m^2.24$ hrs, which is the amount of water passing through unit area in 24 hours in an arbitrary test, but this unit can be converted to moisture vapour resistance by dividing into 207; for example, a polythene vapour barrier may have a moisture vapour transmission of $0.5 g/m^2.24$ hrs which is a moisture vapour resistance of 414 MNs/g, whereas a breather paper may have a moisture vapour transmission rate of $150 g/m^2.24$ hrs which is a moisture vapour resistance of 1.38 MNs/g.

When using a paint on an external masonry surface it is important to ensure that the moisture vapour resistance is lower than the surface to which it is applied, to avoid the danger of interstitial condensation beneath the coating.

9.5 Roof coatings

Various coatings are marketed for the maintenance of slate, tile and sheet roofs. Although asbestos cement slate and sheet roofs can deteriorate and become more permeable with weathering, and their life may be extended by applying a suitable coating to the exposed surfaces, in all other cases applied coatings are far less durable than the surfaces to which they are applied. If tiles are spalling in frost their deterioration would not usually be prevented or even slowed by the application of a coating which may actually encourage condensation to accumulate within the tile and thus encourage spalling damage. The usual problem with roofs is loose slates or tiles, but this problem can only be remedied by stripping and replacing the roof covering; the same slates or tiles can be used if they are in reasonable condition, old slates and tiles often being more durable than modern replacements. Sometimes coatings incorporating glass fibre, nylon or hessian reinforcement are advertised as remedying loose tiles or slates, but they generally result in large sections of the roof covering becoming detached rather than individual tiles and slates, and they certainly do not improve the weather resistance of the roofs to which they are applied. Generally coatings for tile and slate roofs must be considered cosmetic alone, but they usually soon deteriorate and the roof then becomes very unsightly.

Reflective coatings are sometimes used to protect asphalt roof surfaces from sunlight. If asphalt temperatures are excessive, the surface crazes due to loss of volatile components and subsequent shrinkage. The resulting crazing fractures become progressively deeper and encourage local loss of volatile components. Suitable reflective coatings can delay the development of failure in this way, although there is usually eventually sufficient movement in the asphalt to fracture the reflective coating, the dark surface of the exposed asphalt then resulting in localised excessive temperatures and volatile loss, and ultimately deep fissures following the line of the original coating fracture. Reflective coatings are therefore only effective if they are frequently maintained to repair and preferably avoid the development of coating failure in this way. In fact, asphalt roofs can be protected much more satisfactorily using light coloured chips, although chips can only be used where they are protected from wind dislodgement by parapets.

Bitumen and tar coatings are also marketed for application to old felt and asphalt roofs. Products of this type are usually emulsions which are cheaper and less flammable, but they can be washed away by rainfall shortly after application and do not normally contain enough solvent to soften the surfaces to which they are applied sufficiently to establish a good bond. Solvent systems are therefore preferable from a performance point of view and can quite efficiently repair weathered asphalt and felt surfaces, effectively replacing components that have been lost by volatilisation. However, these coatings are not effective for repairing roofs which are leaking because of fractures caused by structural movement or, in the case of felt, through adhesion failure between layers, particularly at joints.

Roof systems and repairs are considered more fully in Chapter 13.

Foundation problem

10.1 Introduction

Foundations must be designed to support the loads from a structure, taking proper account of ground conditions. The design of foundations and the calculations involved will not be considered in detail, but only the diagnosis of foundation defects which may be observed in the course of investigations into defects in buildings.

Normal foundations comprise concrete strips on which the walls are erected, together with prepared foundations to support any solid floors. It is well recognised that strip foundations must have sufficient width to provide an adequate bearing area to support the imposed load on the ground support concerned; the depth of the foundation strip is important in terms of the ability of the foundations to resist local fluctuations in the bearing capacity of the ground without developing fractures, and the total depth of the foundations is important if it is necessary to excavate sufficiently deeply to provide bearings on better compacted soil or perhaps even rock at a greater depth. The importance of careful preparation of foundations for solid floors is not so widely recognised, as floor loads are much less and the requirements much less severe, but problems commonly arise if the floor slab support is poorly compacted, or support varies on a sloping site between excavated ground at one end and inadequately compacted fill at the other. If the load bearing capacity of the ground is poor, construction on a substantial reinforced concrete raft may provide the best means for spreading the foundation load, and various piling systems may be preferred to excavating deep foundation trenches.

10.2 Settlement and subsidence

The differences between settlement and subsidence failures are not widely recognised, although the different causes are obviously important in relation to remedial works. The causes may also be critical should damage result in a dispute for negligence or breach of contract in design or construction.

Settlement is due to normal compaction of the supporting ground as the building loads are imposed on the foundations. Some settlement always occurs during construction, but there is usually some further settlement or creep following completion. Settlement is normal and must be anticipated in design, so that damage due to settlement is an indication of inadequate design in relation to ground conditions or failure to observe the design during construction. General settlement usually occurs on loose ground such as sand or shingle or on readily compressed ground with a high organic and moisture content, such as peat. It is only troublesome because the external walls, the internal walls and solid floors impose different loads and therefore settle to a different extent, the most obvious damage being doming or an increase in height of the floors, although actually it is the walls that are settling around the floors, or settlement of the heavily loaded peripheral walls in relation to the internal partition walls. In one extensive housing development on sandy soil in the south of England, extensive damage was caused by the preferential settlement of the partition walls, and this was found to be due to inadequate foundations. Inadequately compacted landfill is also a serious problem, particularly if it contains biodegradable materials, such as wood, which will cause further settlement over a prolonged period. Certainly the worst settlement problems are associated with sites in which the load bearing capacity of the ground varies so that part of the foundations settle, sometimes causing massive structural fractures. In extreme cases, diagonal fractures may be identified as resulting from the settlement of a distinct part of the structure; in other cases fractures, generally through openings which represent the weakest parts of walls, will be wider at the tops of the walls than at the bottoms, indicating that local settlement has occurred to one side of the fracture, such damage being common on inadequately compacted fill, on land with patches of fill, or on sloping sites where walls are on excavated ground at one end but fill or inadequately compacted ground at the other. Once settlement has been fully relieved a damaged building may become structurally stable and no remedial works are necessary, other than the repair of the settlement fractures, unless structural weakening has occurred which requires the fractures to be rebonded to restore structural integrity, as when vertical fractures occur between gables and adjacent bracing walls. However, if foundations are seriously inadequate and settlement is continuing, it may be necessary to provide additional support by inserting deeper foundations or perhaps supporting piles.

A special situation arises when buildings are constructed on shrinkable clay soils. Normally clay soils will be moist and in their expanded condition, but abnormal drying may result in shrinkage. Such problems are not usual in the British Isles, but serious damage occurred in this way through the exceptionally dry weather in 1976. In addition, many of the houses constructed on shrinkable clay during the exceptionally dry weather have since suffered damage due to expansion of the clay on subsequent wetting. Buildings themselves and associated patios, paths, drives and roads all reduce rain penetration into the ground and can result in clay shrinkage damage following construction. For these reasons deeper foundations must always be used on clay soils to provide support below the clay or at a depth within the clay which will not be affected by such moisture content changes.

Subsidence follows from some unexpected event following construction. Ground water percolation may result in removal of material and loss of support. Water percolation can occur in this way through natural groundwater movement, but usually some event prompts subsidence, such as the diversion of a stream, the overflow of a drainage system, or even a fracture in the rainwater or sewage drains associated with the building itself, although one of the most common causes of subsidence is the fracture of a water main due to frost or traffic damage. Landslip is also another example of subsidence, usually because apparently stable ground has been fluidised through the accumulation of an exceptional amount of rainwater, sometimes causing severe damage to houses and gardens constructed on sloping sites.

10.3 Tree root damage

Tree root damage results most obviously from the penetration of tree roots into masonry and beneath foundations, and rupturing due to progressive root growth. Such damage can usually be readily identified by excavation and does not justify special comment, except in relation to safe separation between trees and buildings.

A more serious problem is the presence of trees in conjunction with shrinkable clay. Deciduous trees will remove water from the clay during the summer and the clay will shrink, but the tree will have no demand for water during the winter so that the clay moisture content will increase and it will expand. This seasonal movement can only be avoided by ensuring that buildings are constructed a sufficient distance from established trees, or new trees are planted a sufficient distance from a building. If these requirements cannot be satisfied deeper foundations are necessary to penetrate below the clay, or sufficiently deeply in the clay for moisture content fluctuations to be minimal. If a new building is constructed in the summer on a site from which trees have been recently removed, clay soil may have an unusually low moisture content; subsequent wetting may cause expansion and damage to the new building, unless the foundations are sufficiently deep to avoid the effect as previously explained.

Safe separations between trees and buildings depend upon the tree species and are summarised in Table 10.1.

Table 10.1 Recommended minimum distances between trees and buildings

Tree species	Minimum separation on shrinkable clay soil	Normal maximum tree height (H) (m)
Oak	$1H$	16–23
Poplar	$1H$	24
Lime	$\frac{1}{2}H$	16–24
Ash	$\frac{1}{2}H$	23
Plane	$\frac{1}{2}H$	25–30
Willow	$1H$	15
Elm	$\frac{1}{2}H$	20–25
Hawthorn	$\frac{1}{2}H$	10
Maple/sycamore	$\frac{1}{2}H$	17–24
Cherry/plum	$1H$	8
Beech	$\frac{1}{2}H$	20
Birch	$\frac{1}{2}H$	12–14
White beam/rowan	$1H$	8–12
Cypress	$\frac{1}{2}H$	18–25

11

Wall problems

11.1 Introduction

The main function of the wall of a building is to isolate the interior from the exterior conditions. Just the presence of the wall is sufficient to achieve this function reasonably, as is demonstrated by the efficiency of a well designed tent. However, the efficiency of walls can be greatly improved in relation to excluding dampness and thermally insulating the interior.

A second function of a wall is structural and to support the load imposed by any suspended floors and the roof above. These basic functions have to be appreciated when designing walls to achieve a certain aesthetic effect, but these various requirements may conflict and may result in various wall problems. Most problems have already been considered in detail in earlier chapters, so that this chapter is essentially a review drawing attention to appropriate matters mentioned earlier, together with comments on actual defects and failures that have been observed.

11.2 Masonry

It has already been explained in section 4.8 that a normal porous solid masonry wall prevents rain penetration by acting as a reservoir which absorbs incident rain and holds it, without allowing it to penetrate through to the interior, subsequently dispersing the absorbed rain by evaporation during dry weather. The interior of a building will remain dry if the average rate of evaporation can exceed the average rate of absorption, provided there are no abnormal absorption peaks, although penetration is often apparent where walls are thinner, such as at window reveals. Problems usually arise because this function of a porous masonry wall is not understood. Construction in impermeable stone may lead to serious problems, particularly as water flowing down the outside surface will be absorbed only into the mortar joints with only limited absorptive capacity, and there is therefore a danger of penetration to the interior. In fact, old solid walls constructed from impermeable stone, such as granite, often comprise two skins with rubble between, and penetration through a mortar joint in the outer skin may result in dampness becoming apparent on the inner skin some distance away and usually lower down the wall, so that the original source of the penetration is difficult to identify. If water penetration is coming through a solid wall, the best remedy is usually to ensure that the mortar joints are sound and well pointed, and perhaps to apply a silicone resin water repellent as described in section 4.11, treatment on impermeable stone walls reducing water penetration by protecting the joints alone.

The thermal insulation value of a solid wall increases with thickness but reduces with density. There are obvious limitations to the efficiency of a solid wall in terms of both resistance to penetrating rainfall and thermal insulation, but cavity construction achieves the most dramatic improvement in both these properties. Cavity problems are discussed in more detail later in this chapter in section 11.5.

Thin cladding is an alternative to cavity construction. In cavity construction the two leafs are self supporting, but cladding is supported by the main wall or inner leaf. The weight of the cladding is usually supported on corbels, which are attached to the background and usually fit into slots in the rear of the cladding panels, the separation of the panels from the background and the junctions between the panels usually being formed using dowels, S-hooks and tie-back cramps. The normal method of erection is to support a panel on corbels or on the panel beneath, roughly locate the cramps in holes in the background filled with mortar, and push the panel back into position against dabs of cement on the surface of the background. When the fixings are firm the joints must be filled, a difficult job on porous stone as they are so fine, usually the width of the wire used to form the fixings, although on impermeable granite and marble cladding, a smooth mortar prepared from cement and fine sand can simply be rubbed into the joints and wiped off the surface.

Cladding problems generally arise through carelessness in design or workmanship, particularly the preparation and erection of the panels. In relation to wind effects, panels only require the same strength as normal single glazing and, with most stones with a reasonable modulus of rupture, quite thin panels would be perfectly acceptable if their edges were continuously supported, as with single glazing in a normal window frame. In fact, conventional fixing systems impose very heavy loads on corbels housed in slots in the backs of panels, and it is necessary to provide a sufficient thickness of panel to provide an adequate bearing surface to support the load. Dowels, S-hooks and tie-back cramps are housed in holes drilled in the edges of the panels which are subject to relatively high stress when panels are affected by wind forces, and again the panel must be sufficiently thick to withstand these stresses. These conventional fixing systems are thoroughly unsatisfactory, resulting in the use of excessively thick panels, and fixing failures account for most of the problems with cladding. Various continuous edge support systems have been introduced, such as the Zibell system, but they have never been adopted to a significant extent in the British Isles, although much more sophisticated fixing systems are used in many other countries. Common cladding failures therefore usually involve poor location of panels during fixing – the obvious sign of an inexperienced fixer – damaged fixing holes in the edges of panels, fracture plains introduced through careless handling during manufacture, transport or fixing, or horizontal fractures due to temperature effects when cutting corbel slots too rapidly, and pieces detached from panels during careless handling. Inherent fractures, that is stress weaknesses which have been introduced through careless working or handling but which have not resulted in actual fractures and detachment, usually because of crystal structure bonding between the broken sections, can be a very serious problem in exposed buildings as wind effects can cause the panels to fracture, perhaps resulting in pieces of panel becoming detached and falling from the building. For this reason it is now a normal requirement that cladding for buildings in London should be checked for fine fractures before erection to avoid these dangers. Stresses in cladding due to differential structural movements can also cause serious problems; they are always related to the use of very rich and inflexible jointing materials coupled with inadequate movement joints, and they will be discussed later in this chapter, in section 11.6.

Some years ago a series of failures developed in marble and granite cladding on major buildings, mainly in the London area. It was observed that the panels were actually bulging outwards until eventually they fractured explosively, scattering sharp pieces of stone over a wide area. The bulges in the panels were located over the original fixing dabs which are used to restrain the movement of the cladding panels as they are pushed back in position and the mortar around the tie-back cramps can set to retain their position. Dabs should be prepared using a rich mix of ordinary Portland cement and a fine aggregate, as this gives an appropriately stiff and very smooth consistency, the dab shrinking after it sets to leave a small gap between the surface of the dab and the rear of the cladding panel. However, many fixers are paid by piece work and are not willing to take the time or trouble to properly mix mortar in this way, using neat cement instead. If the space between the panel and the backing is not too great this is perfectly satisfactory, but problems arise with wider separations, either by design but more usually through carelessness in the construction of the rein-forced concrete supporting framework. Experienced fixers use appropriate mortar mixes to overcome this problem but careless fixers use old cement bags, pieces of polystyrene and various other rubbish stuffed behind the panels to support neat cement dabs and prevent them from sagging. Alternatively some fixers thicken the cement by adding plaster, a technique that is perfectly satisfactory at the time of fixing but disastrous later as sulphate attack of the dabs occurs through the sulphate in the plaster slowly reacting with the cement, causing the dabs to expand, and bulging the cladding panels with the catastrophic results previously described.

These comments on masonry cladding have been concerned largely with natural stone cladding, the thickness of the panels normally varying with the modulus of rupture of the stone involved, so that dense and cohesive marbles and granites are generally used where thin claddings are required. The same techniques can be used with cast stone and even with some manufactured conglomerate materials which can be particularly attractive; these conglomerates are usually prepared using carefully graded stone mixed with a cement or resin binder system to form a large block which is subsequently worked in the same way as solid stone. Concrete cladding is also used but usually in a completely different way. Large cast panels are usually supported on corbels or knibs projecting from floor beams and fixed to the frame-work of the building by bolting, with joints between the panels filled with mastic. Such heavy panels are difficult to control during erection and, if they are being handled too rapidly after casting, corners are easily damaged. In such circumstances contractors are keen to avoid the need to obtain a replacement panel, and they will prefer to use an adhesive to attach the damaged piece. Epoxy resin adhesives are often used for this purpose; these are rather unreliable when used on fresh concrete with a relatively high moisture content, and pieces often become detached after a few seasons of thermal stress due to sun alternating with rain. Fractured pieces of concrete may become detached following periods of frost, often causing serious concern regarding the structural safety of the building. If it is necessary to secure pieces of detached concrete, adhesive must never be used alone, but only in con-junction with bronze dowels.

11.3 Render, plaster and tiles

A render normally functions by increasing the thickness of a porous wall, reducing rain penetration by increasing the absorptive capacity, as well as improving thermal insulation. A frequent error with renders is to apply an excessively rich mix, perhaps also containing a waterproofing additive, in an attempt to obtain an imper-meable render. Rich mixes are subject to excessive shrinkage which often results in the development of crazing fractures and detachment from the supporting back-ground, these shrinkage defects developing particularly in renders prepared using excessively fine aggregate, as the water content of the mix must then be increased to obtain workability and the increased water to cement ratio increases shrinkage. An impermeable layer on the outside of a building also encourages the accumulation of interstitial condensation and bossing or detachment of the render in severe freezing conditions. These failures are described in more detail in sections 4.7 and 7.4, the general recommendation being that the mix should be no stronger and preferably weaker than the background, a requirement that applies equally to successive layers of render.

Renders applied to brickwork backgrounds in which the bricks contain excessive amounts of sulphate often fail through sulphate attack. Sulphate causes expansion in ordinary Portland cement, followed by loss of cohesion. Typical failure involves the development of mainly horizontal but some vertical fractures in the render, coinciding with brickwork mortar joints due to the expansion of the mortar, as well as some loss of adhesion of the render to the brickwork. Damage varies greatly in intensity, the most severe damage usually being associated with brickwork that was saturated with rainwater during construction. If bricks are known to have a high sulphate content damage can be avoided by the use of sulphate resisting cement in the brickwork mortar and preferably also in the render. If damage has developed, replacement of both the render and the brickwork may be necessary. Many popular bricks, including Flettons from certain sources and many Scottish bricks, often contain sulphates in sufficient amounts to cause sulphate attack if they become sufficiently wet for sulphate to leach into the mortar and render. The sulphate content of bricks should always be considered and sulphate resisting cement always used if excessive sulphate contents are likely. Sulphate attack problems are described more fully in section 4.4.

The render bond to the background is particularly sensitive to the presence of excessive moisture. The danger of frost damage to the bond due to the accumulation of interstitial condensation trapped behind an excessively dense render has already been described, but water accumulations can also develop in other ways. The modern student accommodation blocks at one of the Scottish universities have cavity brickwork walls finished with dry dash harling or render with parapets protected by cast concrete copings. Following construction of the parapet brickwork, the roof felt was dressed over the parapets as a damp-proof course and trimmed flush with the outer face of the brickwork. The copings were then bedded on top of this felt course and the harling dressed up underneath the outer projection of the copings. Before completion of construction it was observed that the harling was stained beneath each coping joint, indicating that water was penetrating through the open joint. This penetration is normal as thermal movement in copings means that it is impossible to avoid fractures in the joints, but water penetration does not usually present a problem provided that an adequate damp-proof course is installed beneath the copings. In this particular case the course had been trimmed flush with the external face of the brickwork, so that the top edge of the harling was unprotected. Over the years each period of severe frost caused freezing of water that had accumulated behind the harling, disrupting the bond to the brickwork and eventually resulting in detachment of the harling beneath each coping joint.

Lime mortar mixes were originally used for internal plastering, but gypsum plasters have been used almost exclusively in the British Isles for many years. Gypsum, anhydrous calcium sulphate, solidifies following mixing with water by absorption of water of crystallisation. Solidification is very rapid and usually retarders are added to commercial products to make the plaster easier to work. Various aggregates are also added, including lightweight aggregates. Proprietary formulations vary depending on the purpose of the product, particularly the type of background to which it is applied, whether it is intended to perform as an intermediate coat or finish, and whether it is required to provide thermal insulation.

Most problems with gypsum plasters can be attributed to a lack of understanding of their mode of action or insufficient experience in their use. Poor workmanship is usually indicated visibly by an irregular finish. Sometimes the finish appears to be regular but powdery; this is due to excessive wetting of the surface of the plaster to work it to a smooth finish, the wetting removing calcium sulphate binder and thus weakening the surface. Some plasterers are tempted to wet out mixes progressively to maintain workability, but this also causes weakening; gypsum plasters should be prepared in small batches which can be used within a short time. Plasters are designed to ensure adequate bonding; finish grades should only be used over appropriate undercoat grades, and undercoats should be selected in relation to the background to which they are applied. An undercoat plaster is usually described as browning, and a normal grade of browning is suitable for solid backgrounds with low to moderate suction and adequate mechanical key. Special brownings are available for use on solid backgrounds with very high or very low suction, an undercoat for use on a low suction background being described as a bonding plaster. Undercoat plasters for use on metal lathing are also available, as well as finishes which are particularly designed for use on plasterboard. Incompatibility with the background is the most frequent cause of bonding failure, such as the selection of an unsuitable undercoat in relation to the solid background, or the mixing of undercoat systems. Where plaster is applied to a very absorbent background with high suction, wetting the background will assist in killing the suction and encouraging reliable bonding, as otherwise the bond will be weakened by the loss of calcium sulphate binder through powerful absorption of water from the plaster into the background.

Gypsum plasters are very sensitive to dampness, progressively softening and blistering, and they should never be used in situations where there is a risk of dampness, even following damp-proofing treatment. Many of the specialist contractors involved in remedial treatments for rising dampness suggest that lightweight aggregate gypsum plasters are particularly sensitive to dampness but this is not true; the rate of failure varies with different gypsum plasters, but all of them are susceptible to dampness damage. These problems are best avoided by using a moisture resistant undercoat plaster, such as a mix of cement and sand with a waterproofing additive or a proprietary product; renovating plasters are suitable, except those based on gypsum. Bitumen and tar waterproofing membranes can be applied to walls to similarly achieve damp-proofing but they must be blinded with sand whilst tacky to provide an adequate key for subsequent plaster.

The natural moisture in gypsum plasters is usually sufficient to cause corrosion of iron and steel. Shadow lines on painted plaster walls, usually eventually darkening to a distinct brown colour and sometimes also being apparent through wallpaper, are usually associated with corrosion of steel conduits or wiring covers. Similar corrosion is sometimes associated with plastering beads used to give a sharp finish to edges; stainless steel beads should always be used to minimise this problem, as even galvanised beads often rust where the galvanising is damaged during plastering.

Problems with wall tiling are generally concerned with poor bonding to a substrate, usually through applying tile adhesive to an excessive area so that it partly sets before the tiles are placed. However, tiling also sometimes suffers from movement problems. Tiles are very stable, new tiles expanding only slightly as they reach equilibrium with atmospheric moisture, but they are often applied with dense inflexible joints between them to backgrounds that suffer much greater movement. The most serious problems arise when tiles are applied, perhaps over an intermediate undercoat of sand and cement or gypsum plaster, to solid cast concrete walls which commonly occur around lift shafts and stair wells in major buildings. Concrete shrinkage can cause compression of the tiles between internal corners and enormous stress can be generated, the tiles eventually becoming detached with almost explosive force. Good practice therefore requires movement joints be provided through tiling at all internal corners, as well as at intervals in long runs of tiling; generally the top edge of the tiling on a wall remains unrestrained, perhaps concealed behind a suspended ceiling, to allow similarly for vertical movement, although horizontal movement joints will be necessary with substantial unobstructed heights of tiling. Problems usually arise because vertical movement joints have not been provided, or they do not pass through the full depth of the tiles and the background plaster to the structure behind. In one very large office development in Dublin, wall tiling defects occurred in this way in most of the 52 toilet suites, clearly indicating the consistency and inevitability of such problems in the absence of appropriate precautions.

Another case involved tiling to seven bathrooms and toilets in a substantial domestic property. The backgrounds varied from old external solid rubble walls and old brickwork partition walls, to new concrete blockwork and stud partition walls. Tile adhesion was excellent in all cases, but on stud partition walls tiles fractured at projecting ends and vertical fractures also developed at intervals. The vertical fractures occurred at the joints between the background plywood sheets with convex distortion of the sheets between the vertical fractures. These problems were caused by shrinkage of the plywood, which curved due to the powerful adhesion of the inflexible tiles, the problem being primarily due to a failure to fix the plywood sheets to intermediate studs to resist the development of this curvature. With projecting stud partition walls, the damage to the end tiles was caused by simple differential shrinkage of the plywood background in relation to the stable tile surface.

External walls with poor thermal insulation properties encourage condensation which may result in the development of mould fungi. Minute beetles, often described as Plaster beetles, may be seen which browse on this mould. The families most commonly involved are the Cryptophagidae and the Lathridiidae.

11.4 Timber frame

The literature contains dire warnings of the dangers associated with constructing buildings in timber frame rather than in conventional masonry construction, but these warnings are grossly misleading. In all forms of construction it is obviously necessary to understand properly the materials involved, and problems with timber frame construction only arise through the reluctance of designers and contractors to recognise the differences between timber and masonry. In some respects timber frame construction is extremely advantageous, particularly in convenience and speed of construction when appropriately experienced contractors are using building systems involving extensive prefabrication. Because of its lower density, wood is generally more efficient than masonry in preventing heat loss, but less efficient in controlling noise transmission. However, the main problems arise in regard to differential movement and the susceptibility of wood to decay.

Wood decay problems and their avoidance by preservation have already been discussed in detail in sections 6.4 and 6.7. Obviously sole plates are particularly at risk and must always be preserved, but the risk to normal frame walls is not usually recognised. In standard timber frame construction systems, external sheathing plywood is usually protected from any rain or snow penetration through the external weather skin by a suitable breather paper, but the risk of decay is actually associated particularly with interstitial condensation, caused by humid air from the interior of the building diffusing through the structure and condensing as it approaches the cold exterior, as explained in more detail in sections 3.3 and 4.7. An impermeable vapour barrier, usually polythene sheet, should be provided on the warm interior side of the main insulation; the vapour barrier is usually positioned between a plasterboard interior lining and a glass fibre insulation quilt filling the cavities within the timber frame. Problems arise because this system is critically dependent on the integrity of the vapour barrier, and if the barrier is only slightly damaged during construction, or deliberately damaged perhaps by electricians, sufficient vapour will diffuse through the damage to cause dangerous accumulations of interstitial condensation after several years. Wood should always be adequately preserved to provide protection against possible structural collapse due to decay, although it is not generally recognised that the decay risk is greatest in the external plywood sheathing which should be manufactured from naturally resistant or preserved wood; fungal decay resistance of plywood is defined in British Standard specification BS 6566:Part 7:1985 *Plywood, Part 7, Specification for classification of resistance to fungal decay and wood borer attack.*

Thermal movement in wood is minimal, but moisture movement is usually considerable across the grain although minimal along the grain. In platform construction the walls are erected on top of the floor platforms, so that the main cross-grain movement develops across the floor joists, as well as across the sole plates and the top and bottom plates in each wall section. Generally the movement comprises shrinkage following construction, and it is generally considered that allowance should be made for shrinkage of about 5 mm per storey. If a timber frame building is clad in clay brickwork it is also necessary to allow for brickwork expansion of about 2.5 mm per storey. It is therefore generally recommended that a movement gap should be allowed at the top of brickwork cladding, the gap being 10 mm for one-storey, 15 mm for two-storey and 20 mm for three-storey construction. If this movement gap at the wall head is not provided, there is a danger that eaves soffits and even rafters may be distorted and dislodged. Similar provision for movement is necessary around all window openings, a point that is often neglected. The most common problem is careless construction, particularly the pointing of the wallhead gap, which should be left open to permit movement.

Brickwork and blockwork cladding must be tied to the timber frame. Construction in the British Isles has generally followed the North American practice in this respect for many years, using thin flexible strip ties bent to an L shape with the vertical arm nailed to the timber frame and the horizontal arm embedded in the masonry. Whilst the flexibility of these ties is desirable to accommodate any differential movement, nailing only at the top end of the vertical arm means that the ties are free to flex and are therefore ineffective in restraining the separation between the timber frame and the cladding. Improved ties have now been introduced which are fixed to the timber frame close to the inner end of a horizontal tie bar, although it will be appreciated that they must be attached to the vertical studs to ensure a firm attachment. It is therefore normal to mark the lines of the studs on the breather paper fixed to the surface of the external sheathing plywood, although buildings have been seen in which the studs are not marked in this way, and also buildings in which ties are not fixed to studs even though proper markings are provided!

Although brickwork and rendered blockwork cladding is often used in the British Isles, so that timber frame buildings resemble conventional masonry buildings, many simpler forms of effective cladding are available. The most widely used claddings are tongued-and-grooved vertical cedar boarding, cedar shingles, clay tiles, and asbestos-cement or similar sheet tiles. All these systems are fixed to horizontal battens, although it is important to appreciate that vertical counterbattens should first be fixed to the breather paper covered sheathing plywood to permit ventilation of the breather paper and dispersal of any interstitial condensation accumulating within the plywood. Horizontal shiplap boarding can be applied to vertical battens; western red cedar is usually used in the British Isles but European redwood (Scots pine) impregnated with copper-chromium-arsenic preservative is equally suitable and maintenance free if the greenish-grey weathered coloration is acceptable.

11.5 Wall cavities

Wall cavities were originally vented at the top and bottom to disperse any moisture accumulating in the cavity due to rain penetration through the external leaf or condensation resulting from diffusion of humid air through the internal leaf. However, in an attempt to reduce unnecessary heat loss unventilated cavities were adopted, but serious problems arose. The most dramatic problems were associated with unventilated wall cavities linked with unventilated spaces beneath flat roofs: rain penetration through the external wall leafs caused the diffusion of high humidity air into the roof spaces, with condensation beneath the cold roof coverings often resulting in sufficient accumulations of water to cause severe dripping onto the ceilings beneath, as well as extensive fungal decay development. In fact, normal cavity ventilation does not have a substantial effect on thermal insulation value. Thermal calculations for a typical cavity wall in section 3.2 suggest a total resistance or R value of 1.009 or thermal transmittance or U value of 0.991, assuming that the cavity is unventilated. A freely ventilated cavity would reduce the R value to 0.949, giving a U value of 1.054, an increase in heat loss of only about 6%. In fact, normal wall cavity ventilation is limited and the effect is much smaller, typically only a 1 or 2% increase in heat loss through the walls, which is insignificant in relation to the total heat loss from a building; the ventilation only needs to be sufficient to permit diffusion of humid air from the cavity to the exterior, the high water vapour pressure of humid air ensuring adequate diffusion even with limited ventilation.

Heat transfer across a cavity is due mainly to convection movement of the cavity air. Suitable cavity fill will reduce this movement and therefore reduce heat loss, but the use of unsuitable fill may allow rainwater to penetrate across the cavity. Closed pore foam materials, such as some of the polyurethane foams, are efficient in this respect, although they are impermeable and may result in accumulations of interstitial condensation within the inner leaf, particularly if the inner leaf has good thermal or insulating properties, such as aerated concrete blocks. It is therefore more sensible to use a loose cavity fill which can be thermally as efficient, whilst still allowing some limited diffusion and dispersal of water vapour across the cavity. The most efficient material is granulated expanded polystyrene. Mineral or glass wool insulation is most extensively used but, although it is meant to be treated with a water repellent to prevent it from conducting moisture across cavities, it is often found to be saturated with water, sometimes causing significant dampness problems. Whilst it is possible that this problem is due to inadequate water repellent treatment of the wool at the time of manufacture, it is more likely that the high humidity of the cavity encourages biodeterioration of the water repellent. Such problems are particularly severe in walls subject to sea spray, as the salt tends to overcome the water repellent treatment. In new construction cavity insulation is usually introduced as sheets of glass fibre or foam material, fixed to the inner leaf with a narrow cavity between the insulation and the outer leaf.

Ties must be provided across external wall cavities to ensure the stability of the separate leafs. Various forms of tie are available, the most popular being the butterfly wire ties and the fish tail strip or bar ties. Increasing problems have been encountered over the years through corrosion of galvanised ties, apparently because fresh mortar removes the protective galvanising; the exposed steel remains protected by the alkalinity of the mortar, but eventual carbonation neutralises this alkalinity and enables corrosion to occur. Heavier galvanising has been specified, although bitumen and particularly tar coated ties are more durable, and stainless steel ties are generally completely reliable.

Mortar slovens falling into the cavity during construction can accumulate on ties to form bridges between the two leafs. These bridges seldom cause any problems in ventilated cavities in which there is sufficient ventilation to cope with any moisture moving through the bridges, but in unventilated and filled cavities drying of the bridges cannot occur in this way and isolated damp patches sometimes appear on the interior plaster and decoration where each cavity tie occurs. Cavity fill insulation should never be installed in walls with significant accumulations of slovens on ties, and installers should always check this point as part of their assessment of suitability of a wall for cavity fill installation. The inner surface of an outer leaf may also erode and spall, particularly in walls exposed to sea spray, and debris may accumulate at the base of the cavity, perhaps bridging the damp-proof course in the inner skin and causing rising dampness. The accumulation of debris in this way must be taken into account when diagnosing apparent rising dampness in cavity walls. Drilling exterior leafs to install cavity fill will produce more debris but also dislodge material loosened by other causes, and may lead to severe rising damp problems of this type.

Water penetrating through the outer leaf into the cavity is prevented from affecting soffits over window and door openings by cavity trays, which should divert the water out through weepholes in the outer leaf. Dampness on the soffit or either side of an opening is often caused by discharge from the tray above, usually because the weepholes have been omitted, but sometimes because the tray has been deformed or filled with mortar slovens during construction, or the outer leaf is very permeable and allows excessive rain penetration. Dampness on either side of an opening can also be caused by carelessly formed cavity closures; the inner leaf should be extended into the cavity to form the reveals, but should be protected from dampness in the outer leaf by a vertical damp-proof course.

11.6 Movement

Structural movement problems have already been considered in sections 2.3, 3.5 and 4.2. Although expansion or contraction may be a problem in itself in large structures, differential movement is usually much more serious, involving differences in movement between adjacent materials. Thermal movement, that is movement caused by changes in temperature, tends to be reasonably constant for masonry materials and steel. Aluminium moves twice as much, uPVC moves four times as much, whilst wood moves only a third or half as much along the grain, but three to five times as much across the grain. It is not therefore surprising that movement problems occur particularly in association with wood, aluminium and uPVC windows installed in masonry.

Most concrete mixes shrink by about 0.08% on initial drying, with subsequent reversible movement of 0.02–0.06%. Wood is virtually stable along the grain, but softwoods move by 0.45–2.6% across the grain and hardwoods by 0.5–4.0% across the grain. This high cross-grain movement for wood has been discussed in detail in section 6.3, as well as in section 11.4; the most serious problems with wood movement arise with brickwork or blockwork clad timber frame construction, and the differential vertical movement between the cladding and the timber frame.

The earlier section 11.2 on masonry included a discussion on thin stone cladding. Differential movement between stone cladding and a supporting concrete frame represents a particularly serious problem, mainly resulting from the initial irreversible shrinkage of the concrete, although there is some subsequent differential movement, usually thermal movement due to excessive heating or cooling of the external cladding. It is therefore essential to include movement joints in cladding, particularly horizontal joints beneath each cladding panel supported by corbels on the structural background. In designing movement joints it is necessary to ensure that they will be sufficiently wide to accommodate anticipated movement; if total differential movement of 3 mm is expected, and the preferred mastic has an extensibility of 20%, a movement joint 15 mm wide will be required. It is also important to allow for tension of the mastic as well as compression, and careful attention must be given to bonding the mastic to the surfaces on either side of the movement joint.

Window problems

12

12.1 Introduction

Window design in construction is a very complex subject and it is not intended to consider it in detail in this book. Instead this chapter reviews problems encountered with windows in the hope that this information will assist in the diagnosis of defects on site.

Glazing is considered to include the system in which glass is supported, thus including double-glazed units, bedding compounds and beads. Frames are considered separately as they involve different problems, and there are short comments in this chapter on patent glazing and winding gear.

12.2 Glazing

Normal glazing glass is one of the most durable materials used in buildings. There are, in fact, only three common situations in which glass can be damaged, other than through physical scratching. If the external leaf of a cavity wall contains an excessive amount of unreacted lime and the leaf is also unusually permeable (the situation that sometimes occurs, for example, when some concrete and cast stone blocks are used), water saturated with lime runs down the cavities and accumulates on the cavity trays over the windows, normally discharging to the exterior through weep holes. In exposed situations the discharge from the weep holes may be blown onto the windows, the unreacted lime causing etching and also 'chalk' accumulations. Etching can also be caused if water washings from lichen infected surfaces come in contact with glass, these washings containing oxalic and other organic acids; these acid washings also cause deterioration of lead gutters, as described in section 13.2. Silicone resin sprays can also bond strongly to glass surfaces, usually resulting in a frosted appearance; if silicone resin water repellents are being spray applied to adjacent masonry, any contamination on the glass must be removed as promptly as possible using a cloth soaked in water and detergent.

Glass fractures in normal service are usually associated with metal frames and occur in two ways. The thermal movement for an aluminium frame is much higher than for glass, so that if the glazing has not been cut strictly square and has snags, local excessive stress may develop in cold weather through the frame contracting around the glass, causing fractures to develop. Rusting of steel frames can result in similar stresses and fractures, although most steel frames are now galvanised to avoid such corrosion problems; unfortunately some frame fixers foolishly use ordinary steel screws which result in the development of localised corrosion expansion. Similar damage can occur in wood window frames through rusting of the steel fixing spriggs that are used to retain the glass while the putty bedding is applied.

Glazing compounds can be conveniently divided into two groups, those requiring protection by paint, and those that do not require protection. The first group include traditional setting putty, as well as some types of non-setting mastic. Traditional linseed oil putty is used for bedding glass into wood surrounds. It is usually considered that it can be used on primed wood, although it relies on the absorption of some of the oil into the wood to achieve reliable drying and solidification. The putty progressively oxidises and shrinks so that it progressively fractures and separates from the glass. Failure in this way can be delayed by painting the putty as soon as it has acquired an adequate skin, ensuring that the paint extends well beyond the putty onto the glass and that it is subsequently carefully maintained.

Metal casement putty is formulated differently and will solidify by absorption of oxygen, in the same way as normal alkyd paint systems. This type of putty is intended for use on metal or non-absorbent wood frames; it is actually suitable for use on all wood frames provided that the rebates are painted before applying putty. Again, painting of the finished bedding is necessary to ensure reliable durability, but particularly to ensure good contact with the glass.

The problem of poor adhesion between putty and glass can be avoided to a certain extent by using a non-setting mastic, although this is not extensively used for this purpose.

Amongst the glazing compounds that do not require painting, non-setting compounds are usually used to line the rebate and to cover the surface of the glass before securing the glass using a glazing bead. Setting sealants can also be used and may generate sufficient strength to retain the glass without the use of a bead.

Shaped strips of set material, usually described as rubber or plastic, can also be used in conjunction with glazing beads, or alternatively as gaskets fitted into appropriate slots in extruded frames. Strip materials of this type usually fail at the corners; indeed, strip materials are often used without joints at the corners, allowing water penetration as well as wind penetration which can generate unacceptable noise.

Double glazing, in which two sheets of glass are separated by a gap or cavity, provides improved heat and noise insulation as previously explained in Chapters 3 and 5. Double glazing can be achieved in various ways, perhaps the simplest being secondary double glazing in which an additional light frame is attached to the inner side of an existing single-glazed window frame. One serious problem with secondary double glazing is the high risk of condensation. It will be appreciated from the discussion in section 4.7 that the risk of condensation is related to the humidity of the air and thus its dew point, as well as the temperature of the surfaces with which it is in contact, condensation inevitably developing on surfaces with temperatures lower than the dew point. One problem with secondary double glazing is that, when the secondary glazing is closed, normal room air is trapped within the double glazing space, this air having the same dew point as the room air. Unfortunately, one of the effects of secondary double glazing is to substantially reduce the temperature of the inner surface of the original single glazing, thus greatly increasing the risk of condensation. Thermal resistance or R values can be calculated for each layer in an element and these can be added as explained in section 3.2 to give the total R value, the reciprocal of thermal transmittance or U value. However, the temperature drop across each layer is proportional to its resistance, so that resistance calculations can enable temperature to be determined at each interface. In making these calculations it is, of course, essential to take account of surface resistance, so that the inner surface of glazing is always cooler than the air with which it is in contact. If it is assumed that the internal air in a room has a temperature of 20 °C and the external air has a temperature of 0 °C during winter conditions, interpretation of the thermal calculations for single glazing in section 3.2 shows that the temperature of the internal surface of the glass is 6.4 °C. If the calculations are then repeated for secondary double glazing with an air gap in excess of 20 mm, the temperature of the secondary glazing surface in contact with the room air is 13.2 °C, considerably warmer because of the improved thermal insulation value of the double glazing and therefore reducing the risk of condensation on this surface. However, the inner surface of the original single glazing is now isolated from the warmth of the room and has a temperature of only 3.2 °C, so that the risk of condensation on the inner surface of the original glazing is enormously increased by the installation of the secondary double glazing. Manufacturers of secondary double glazing do not usually appreciate these dangers and often make grossly misleading claims in relation to the condensation reducing efficiency of their system. In fact, the danger of condensation within the air gap can only be reduced by ensuring that both the original windows and the secondary double glazing are efficiently sealed, and then providing a moisture absorbent within the air gap to reduce the humidity and dew point of the trapped air.

This is precisely the system that is normally adopted in manufactured hermetically sealed double glazing units. Manufacture usually involves the attachment of two pieces of glass to edge spacers using a suitable adhesive. Ideally the edge spacers, in combination with the adhesive, form a reliable and permanent seal. The humidity within the air space is reduced to a minimum to reduce the risk of condensation by manufacturing units in an atmosphere of dry air or inert gas, or by providing a reservoir of absorbent within the edge spacers. The edges of the units are usually finished with a tape, typically with an aluminium foil surface, to protect the edge spacers and to make handling easier, but also to screen the assembly adhesive from ultra-violet light to extend the durability of the seal system. Many manufacturers have discovered that such seals are unreliable and they have filled the gap between the glass outside the spacers with mastic to improve sealing, some manufacturers actually using spacer adhesive only as an assembly aid and relying entirely on this mastic for sealing. Systems of this type can have an extremely high failure rate if they are carelessly assembled, as the slightest gap in the mastic sealant will allow air penetration. Some manufacturers have also found that a vacuum gap is advantageous, particularly in the way in which it assists assembly, but again the system is only as reliable as the seals, and an effective vacuum also distorts the glass and vision.

Manufactured double glazing units are fitted in frame rebates in the same way as single glazing, except that the rebates must be sufficiently deep to accommodate the increased thickness. Obviously heavy and cumbersome frames are not very attractive and most early systems had small air gaps, typically only 5 mm, although a gap approaching 20 mm is necessary to achieve optimum thermal insulation performance, as is apparent from Figure 3.1 in section 3.2. Purchasers of double glazing are now more sophisticated in their demands and recognise that their expenditure on double glazing must be justified in terms of thermal insulation efficiency, and gaps of 10–15 mm are now more usual.

Stepped double glazing is a method that has been designed to allow double glazing to be fitted in normal single glazing frames. In effect, a second sheet of glass is attached to the surface of single glazing; usually in stepped double glazing the larger sheet of glass fits the normal single glazing rebate, and an attached smaller sheet of glass protrudes through the window aperture. Stepped double glazing is just as efficient as flush edge double glazing with a similar air gap, although it is not particularly attractive. Stepped double glazing is only acceptable, of course, in frames that are sufficiently robust to stand the additional glazing weight without distortion.

In the winter it usually seems that the air inside a building is much drier than the air outside. In fact the outside air is usually driest! This phenomena is due to the temperature difference which means that the interior air is warmer with a much greater water holding capacity, so that a high air moisture content or humidity will only result in modest relative humidity or 'wetness'. In contrast the low exterior temperature will ensure that the air will be saturated with 100% relative humidity by quite low humidity. The significance of this discussion is in the fact that the interior humidity is actually much higher than the exterior humidity, and as water vapour pressure depends on humidity there is a pressure gradient which tends to cause water vapour to vent naturally towards the exterior. This observation prompts the thought that condensation within double glazing could be avoided most efficiently and cheaply by venting the cavity to the exterior, in the same way that wall cavities and roof spaces are vented to avoid interstitial condensation!

12.3 Frames

The thermal properties of window frames vary greatly, and as the frames represent 20–30% of window surface area, they are very significant in relation to the total thermal efficiency of a window. Single glazing has a typical thermal transmittance or U value of about 5.6 for normal exposure, whilst double glazing with a gap of about 20 mm or more has a U value of about 2.7. An ordinary painted steel or aluminium frame has a U value of 5.6, similar to single glazing, heavy painting or polished aluminium surfaces both achieving slight improvements in insulation, but a wood frame in contrast has a typical U value of about 1.7. Thus a single-glazed metal frame window will have a U value of 5.6, whereas a single-glazed wood frame window in which the frame forms 30% of the window area, will have a U value of about 4.3. If a wood frame is double-glazed with an air gap of more than 20 mm, the U value is reduced to 2.5, certainly the most efficient system that can be achieved at reasonable cost. An aluminium frame fitted with manufactured double glazing units with a 6 mm air gap will have a total U value of about 3.8, assuming that the frame forms about 20% of the window area. Increasing the gap to 20 mm and providing a thermal break in the frame will improve insulation efficiency to a U value of about 3.2.

Wood frames have become unpopular, despite their thermal insulation efficiency, because of their poor durability, both in regard to the susceptibility of the wood to fungal decay and the need for labour intensive painting at relatively frequent intervals. Fungal decay can be completely avoided in window frames and other external joinery by appropriate preservation treatment, immersion or double vacuum treatment with organic solvent preservatives being most appropriate as explained in section 6.7; wood coatings are discussed in section 9.2.

Ordinary steel frames usually corrode, even if the glazing compound and the paintwork are carefully maintained to minimise water penetration, rusting often causing fracturing of the glazing as described in the previous section 12.2. The resistance to corrosion can be improved by the use of appropriate coatings or by galvanising, as explained in sections 4.5 and 9.3; most standard steel frames are now galvanised although a few are factory coated with pigmented systems.

Aluminium frames were at one time only available as a polished aluminium finish. Aluminium forms a protective layer of oxide which inhibits further corrosion and ensures reasonable durability, except in heavily polluted or marine conditions, but this oxide formation also dulls the polished surface. Anodic finishing was introduced as a more durable finish for aluminium, although restricted to a brownish or bronze coloration that was not considered appropriate for many decorative schemes. The introduction of unplasticised polyvinyl chloride or uPVC frames with an apparently permanent white finish forced manufacturers of aluminium extrusions and frames to develop white coated systems. Various systems were developed but the most reliable and most durable involved powder coating, that is the application of the material as a powder to the extrusions followed by heat fusion to form the coating. Some proprietary polyester powder systems were particularly effective, but also rather expensive, and cheaper acrylic coatings applied by electrophoretic systems were introduced. These acrylic coatings were less reliable and much less durable, and their use has declined substantially in response to competition from powder coatings and from uPVC frames. These coating systems for aluminium frames are discussed in more detail in section 9.3.

Unplasticised polyvinyl chloride or uPVC frames have been remarkably successful with good thermal efficiency and now reducing costs. Their main advantage over aluminium or wood is their low maintenance cost, largely because their colour is throughout the thickness of the material instead of being dependent on an easily damaged coating. One advantage is the ease with which extrusions can be welded together to form frames, but a disadvantage is their relatively low strength and the need to use quite large sections, often stiffened with steel reinforcement, to achieve the necessary strength. The only frame system that can match uPVC in performance and cost is wood, appropriately preserved and finished with a coating that has fully bonded, but unfortunately such coatings are not readily available and wood frames cannot compete efficiently with uPVC frames at this stage.

12.4 Patent glazing

Patent glazing developed originally as a description for a variety of proprietary roof glazing systems in which the sides of each sheet of glass were secured using special metal glazing bars, basically involving a cap screwed to the bars to clamp the glass but involving, of course, other design features to prevent water penetration. Some modern patent glazing systems are suitable for supporting all four edges of the glass when used in a vertical configuration, and patent glazing was really the stimulus for the development of the various 'dry glazing' systems that are now so widely used in aluminium and uPVC frames. Double glazing systems have been introduced, originally assembled on site, although hermetically sealed double glazing units are now favoured as a means for reducing heat loss through glazed roofs, the lower sheet often being wired glass to provide protection from shattering when exposed to fire.

Most systems are well established and development over many years has ensured freedom from problems, but the comparatively recent introduction of double glazing for patent glazing of roofs has introduced some special problems. The bottom edge of roof patent glazing projects beyond a transom seal, the glazing above the transom being exposed on its underside to the accommodation air, but the glazing beyond or below the transom being exposed to the exterior air. In winter conditions the temperature of the glazing above the transom is therefore higher than the temperature of the projecting section below the transom, although with single glazing the effect is not too great as the entire upper surface of the glazing is exposed to the exterior air. However, where double glazing is involved, the lower inner sheet is isolated from the exterior air and a considerable variation in temperature can then arise between the main area of the inner sheet of glass above the transom and the projection beyond the transom, the main body of the glass expanding to a much greater extent than the projection, which is then put in tension so that fractures may develop. Wired glass is weaker than plain glass in tension and fractures frequently develop when wired glass is used as the lower sheet in patent double glazing, the fractures being almost entirely confined to the projection beyond the transom seal. The risk of damage can be minimised and fractures usually prevented by ensuring that the inner glazing does not project too far beyond the transom seal, a projection of no more than about 25 mm (1") being desirable.

Alternatively, the lower inner sheet of glass can be terminated at the transom seal, only the upper sheet of glass projecting beyond the transom to form a drip.

12.5 Winding gear

A wide variety of proprietary systems are available to open and shut windows remotely. The purpose of this section is not to describe these various systems in detail but simply to comment in general on reliability.

The simplest systems involve levers operated by cords, but they usually rely on tension in the cords to hold the fanlight or other window in the desired position. More reliable control can be achieved if a lever from the opening frame is provided with teeth which engage with a worm gear operated by cords or rods. Another system involves a handle and gears which rotate a vertical rod which is threaded at the upper end and which moves a series of levers to operate the window, a particularly long established and reliable system. However, all these systems involve levers on the opening windows which are often considered to be cumbersome and unsightly, and there have been various attempts to develop more attractive winding gear.

In some systems the levers are replaced by flexible cables which move within covers or pipes, the movement being used to open and shut the windows. Such systems are not much more attractive than lever systems unless they are concealed within the frame and they tend to be less reliable. In modern systems the movement of the cable is used to operate scissor levers which are housed flat against the frame when the windows are closed. In some proprietary systems the cables are replaced with strips of steel running within channels. Both the cable and strip systems can be completely concealed within modern hollow steel, aluminium or uPVC frames with only the operating levers or winders visible. However, whilst these frame concealed systems are extremely attractive, the forces imposed on the scissor levers when commencing opening or when finally sealing the casement into the frame are usually very considerable, and problems arise through wear on the lever pivots, as well as difficulties in adjusting the sliding components to ensure reliable closure. As a general rule systems of this type with winders cause little trouble, but lever operated systems require critical adjustment and are particularly subject to wear.

Roof problems

13

13.1 Introduction

It is not intended that this chapter should consider roof designs in detail, but only aspects of design and construction which result in problems in service. This chapter is therefore divided into sections concerned with problems relating to roof covering materials, insulation, pitch roofs and flat roofs.

13.2 Roof coverings

Pitched roofs are traditionally covered with slates or tiles. Slate rock is formed when successive layers of mud are metamorphosed under heat and pressure to form solid rock, although the layers result in a laminar structure which enables blocks of stone to be cleft into the sheets which are more familiar as roofing slates. Most of the roofing slates in use in the British Isles were quarried in North Wales, although slates of less uniform but perhaps more attractive colour and texture are also available in the Lake District and in Cornwall, where they are used locally. Most roof slates for new construction and repair are obtained from less expensive sources in Spain; most of the Spanish slate is similar in appearance to Welsh slate but it is usually less durable. The long-term durability of slate depends mainly on the nature of the bonding between adjacent laminations, some slate delaminating quite rapidly in polluted urban atmospheres; durability is usually assessed using an acid resistance test. It is a matter of great concern that, when individual slates commence to slip from a roof due to nail corrosion, it is often recommended that the entire roof covering should be replaced, whereas the correct solution to the problem is to carefully remove and refix the original slates which have proved their virtually infinite durability, a replacement covering of any other material being probably less durable, even if it is new slate. Recommendations for complete replacement are usually prompted by the greater labour cost that is involved in carefully removing slates for refixing, although some roofing firms will obviously make more profit from supplying new covering, and some may be particularly attracted by the possibility of both supplying new covering and making extra profit by selling the removed slates which are now very valuable!

Traditional tiles are manufactured from clay, and their durability depends on their porous properties, particularly pore size distribution as explained in section 4.3; tiles are generally more susceptible to frost damage than bricks because of their greater exposure to rainfall and the higher risk that they will be saturated when freezing occurs. Durability is also related, but to a lesser extent, to the cohesive strength of the baked clay, and in tiles from a single clay source it is normal for well fired darker coloured maroon or mauve tiles to be more durable than under-fired pink or orange tiles. Traditional plain tiles have a slightly domed shape which avoids close contact between successive courses, avoiding capillarity between adjacent tiles and allowing them to dry more readily, again improving the durability of the tile roof covering. As good durability is particularly associated with macroporosity or large pore diameters, the most durable tiles also tend to be the most permeable. This permeability is not a problem in plain tiling, as capillarity retains the water within the exposed tile, the previously described doming preventing it from passing into the tile beneath except where they touch; with some modern machine-made plain tiles, however, which lack the doming, and in single lap pantiles and other interlocking tiles, permeability can present a serious problem which tile manufacturers often counter by producing a denser but less durable tile. Tile efficiency in preventing water penetration was well recognised in the past, as penetration was clearly apparent from within the roof, but the current universal use of sarking felt tends to conceal evidence of any inadequacy in the tiling in preventing water penetration. In modern roofs, tile failures are therefore usually limited to spalling damage of the exposed tile surfaces, susceptible tiles often being rather weak when wet and also fracturing, particularly through persons working on a roof to carry out repairs, although fractures sometimes actually occur in this way during construction.

Plain and interlocking concrete tiles are now extensively used. Their properties tend to vary rather more widely than for clay tiles, individual types varying in mix design but individual batches often varying due to poor manufacturing control. Desired colouring is achieved either by adding pigment to the mix or by using coloured sand facing, but neither method is particularly reliable; the bonding of sand to the concrete is usually unpredictable, even in a single batch, and a roof may eventually develop rather unsightly bald patches, whilst pigmentation colours often change or deteriorate with weathering. In addition, pigmented concrete tiles sometimes vary significantly in colour when new, usually due to poor control of water content and compaction during manufacture, a problem that particularly affects small rather than large manufacturers.

Thin asbestos-cement tiles are often used as alternatives to slate; in fact, they are often incorrectly referred to as slates. The asbestos reinforcement allows very thin tiles to be produced which are suitable for lightweight low-cost roofing. Asbestos-cement tiles vary greatly in durability but seldom have an effective life in excess of about 25 years, compared with a virtually indefinite life for good quality slates or plain clay tiles. Today other reinforcements are used in place of asbestos.

Lead is the traditional covering for shallow pitched or flat roofs. Copper and zinc were introduced several hundred years ago but copper is preferred because of its better durability. However, none of these metals are ideal. Copper and particularly zinc corrode in polluted atmospheres, and even lead will corrode if its receives organic acid washings from surfaces supporting heavy lichen or moss growth; tile and stone roofs discharging into lead parapet and valley gutters are particularly troublesome in this respect, although coating the lead regularly with a tar composition will reduce the rate of corrosion. Suitable stainless steel alloys are generally much more reliable and are now being used more extensively; types of stainless steel are discussed in section 4.5.

Asphalt is frequently used for flat roof construction. The softening point of the asphalt must not be too high to avoid the development of brittle fractures during cooling and when subject to subsequent temperature changes, but a lower softening point results in a higher proportion of volatile components, so that exposure to strong sunlight often results in the loss of some components by volatilisation followed by shrinkage and the development of fractures. These problems can be reduced by protecting the surface of the roof from direct solar radiation by covering it with suitable paving or tiles, if it is likely to receive any foot traffic, or with grit, for roofs that are free of traffic and sufficiently sheltered from the wind, perhaps by parapets, to avoid grit scatter. Reflective coatings have also been introduced to reduce asphalt temperatures, but they are not entirely reliable, as explained more fully in section 9.5. Poor bonding at working joints, particularly in forming upturns, account for most reported defects. An unusual defect involved an asphalt covering to the virtually flat top of a mansard-type roof structure, the asphalt terminating at the edges on top of a lead flashing turned down over the top of the tiling of the steep sides of the roof. Three problems were encountered with this detailing. Shallow falls were provided to discharge water over the edges of the roof, but this caused considerable irritation to residents as the water discharged noisily past their windows, and asphalt curbs were set on the roof to divert the water flow to either side of each window. In addition, the asphalt softened at the edges and eventually flowed down over the lead onto the tiling. The asphalt had been correctly selected with a fairly low softening point, to ensure adequate flexibility and resistance to brittle fracture in view of the large dimensions of the roof, but the asphalt had been softened in contact with the lead flashings, which were absorbing solar radiation on their exposed downturns. This problem was eventually solved by cutting back the edges of the roof and reforming them with asphalt with a higher softening point. The third problem with this roof involved the lead downturns themselves; the building was on an exposed coastal site with high wind speeds, and turbulence over the edges of the roofs was sufficient to lift the flashing downturns.

Bituminised felt is often used on shallow pitched or flat roofs as a convenient and less expensive alternative to metal or asphalt. Felt roofs of sufficient thickness constructed from sufficient layers can be very reliable, but it must be appreciated that they are not particularly durable and will require coating or replacement at intervals as part of normal maintenance. The main problem is the development of brittle fracture through the loss of volatile components, as with asphalt, but felt is much thinner and less able to tolerate such fractures. Protection from solar radiation using sand or reflective coverings will improve durability but replacement within 15 to 25 years is inevitable despite such precautions. Deterioration is most rapid on roofs with insulation immediately beneath the felt which isolates the felt from the supporting decking, amplifying the temperature fluctuations affecting the felt and causing premature failure.

Various other roof coverings are sometimes used. Butyl rubber is an effective alternative to felt, although there is little information at present on long-term durability. Butyl rubber was involved in a roof problem following extensive repairs to a Scottish castle which had been damaged by fire. The roof involved was a shallow sloping roof on a large tower within parapets, the slope discharging into a large box gutter. Water leaks affected the ceiling and walls beneath the box gutter, but the source could not be readily identified as no fracture could be seen in the butyl rubber covering. There was no access to the area beneath the box gutter but the author gained access to the adjacent sloping roof, which reduced in height to about 300 mm (12") at a beam running along the edge of the gutter. A small camera with an attached flash was passed between the ceiling joists beneath this beam and into the small space under the gutter; a series of photographs was taken which were later examined to identify the source of water. It was found unexpectedly that the water was leaking through the bottom edge of the gutter on the side adjacent to the roof slope, despite the perfect appearance of the butyl rubber covering. The bottom boards of the gutter were not properly supported in the affected area and people moving in the gutter had apparently caused the split, although it had not been observed despite careful inspection; the moral of this story is that splits in black butyl rubber and other black materials cannot be seen when the material is wet!

Valley and parapet gutters are effectively separate small roofs, usually covered with lead or asphalt. Most problems are associated with careless detailing at discharges, and carelessly formed and inadequately lapped stepped movement joints in lead and copper gutters. Problems with eaves gutters and downpipes are usually relatively obvious, such as sagging and blocked gutters, leaking joints and blocked discharges. Downpipes may also be blocked, and blockages can cause permanent accumulations of water which fracture the pipes through freezing; round cast iron and particularly rectangular cast iron pipes are very susceptible to this form of damage.

13.3 Insulation

A roof structure, that is a roof with the ceiling immediately beneath it, must achieve adequate thermal insulation, but great care is necessary in design to avoid interstitial condensation problems through diffusion of humid air from the accommodation towards the cool roof covering. A traditional pitched roof is usually described as a 'cold roof' construction because the air space between the ceiling and the roof covering is ventilated to the exterior. This ventilation disperses any humid air diffusing through the ceiling from the accommodation and traditional roofs seldom suffer from interstitial condensation problems. However, sarking was generally introduced some years ago, partly to avoid the danger of wind blown rain or snow penetration beneath tiles, but also to reduce roof ventilation to improve thermal insulation; however, condensation often occurs on the sarking felt of roofs of this type, resulting in the development of moulds and wood destroying fungi on contacting rafters and, in extreme cases, causing water drips onto the ceiling beneath and often extensive staining. Fortunately this danger was soon recognised and roof space ventilation is now required in conventional roofs; the ventilation is sometimes achieved through gable, ridge or roof slope vents, but eaves soffit vents are particularly efficient and much more convenient. In this form of construction the required insulation properties are easily achieved by providing a sufficient depth of appropriate insulation on top of the ceiling.

Flat roofs can also be designed with a similar 'cold roof' form of construction, usually with soffit vents providing ventilation to the airspace between the ceiling insulation and the decking supporting the roof covering. However, many flat roofs are not constructed in this way but involve 'warm roof' construction with an unventilated roof space. Humid air from the accommodation can diffuse through the ceiling, roof space and decking, eventually condensing within the decking close to the cold roof covering. If the decking is wood boarding or plywood, or particularly chipboard or strawboard, this condensation can result in the development of wood destroying fungi and the structural collapse of the decking; typically Wet rots develop in the wood in contact with the roof covering which is directly affected by the condensation, but Dry rot can develop further away from the roof covering at intermediate moisture contents and can spread very extensively and very destructively through the roof structure due to the lack of ventilation, as described in more detail in section 6.4.

These interstitial condensation problems in 'warm roof' construction can in theory be avoided by providing an impermeable vapour barrier as close as possible to the ceiling to prevent the humid air migrating into the roof structure. In practice, vapour barriers are inherently unreliable, as humid air may diffuse around the edges of the barrier, such as through walls or through carelessly formed joints, but particularly through damage during construction, such as the insertion of wiring and light fittings. All these faults of a vapour barrier system are recognised but there is a further problem which is not currently acknowledged: there is often sufficient moisture present in the air within the roof space and the wood components to cause decay problems in any case through the moisture concentrating beneath the cold roof covering through interstitial condensation, so that there is a serious risk of decay damage in a 'warm roof' structure, even if the vapour barrier is properly installed and completely intact! Preservation treatment is therefore essential for all wood components and it is also essential to avoid the use immediately beneath the roof covering of any decking or insulation which will be affected by water, such as chipboard, strawboard or insulation board. If decking deterioration occurs through condensation, the roof covering and decking must be completely stripped and replaced with suitable durable materials and, at the same time, the roof must be converted to a more reliable 'cold roof' construction; the roof space must be ventilated and additional insulation must be provided on the ceiling to maintain the required thermal properties.

It is usually considered that these interstitial condensation problems occur mainly in wood flat roofs but they are actually much more widespread than is generally appreciated. Conventional modern pitched roofs with sarking felt often suffer from interstitial condensation beneath the felt if ventilation is inadequate as previously explained, but old lead covered church and cathedral roofs suffer in the same way, the interstitial condensation in the boards beneath the lead usually resulting in Death Watch beetle infestation if the decking is oak or chestnut, this wood borer being dependent on the fungal decay caused by the condensation as explained in more detail in section 6.5. Concrete roofs may also be equally affected. The flat roof of a large new office block was constructed using reinforced concrete beams, with intermediate hollow pots supporting lightweight concrete insulation with felt covering and with a suspended ceiling beneath. The top floor of the building could not be occupied because of water dripping from the ceiling. The contractors suspected that the roof covering was defective but the problem continued despite numerous repairs, as it was actually caused by interstitial condensation beneath the felt roof covering which had saturated the lightweight concrete insulation. The situation was slightly aggravated by humidification of the accommodation air by the air conditioning. Remedial works involved converting the original roof covering to a vapour barrier by adding insulation covered by a new weather membrane.

This is an appropriate point to mention an alternative roof system which is often better known as an 'inverted' roof. The system is conventional in the sense that it involves a decking supporting a felt or asphalt roof covering and a ceiling beneath, but the main insulation is provided on top of the roof covering. Although an 'inverted' roof may be an appropriate remedial solution to interstitial condensation beneath the covering of a 'warm' roof, in the sense that the additional insulation on top will reduce the risk of condensation in this way, it suffers from several disadvantages. Rain and melting snow will percolate through the insulation or between the insulation slabs and cool the roof covering, particularly where channels are left leading to drainage points, so that there is still a risk of local condensation beneath areas affected by drainage in this way. In addition the insulation must be entirely removed if any repairs become necessary to the roof covering, although obviously the insulation will tend to protect the roof covering from weathering and it is more likely that the insulation will need to be replaced.

The obvious conclusion, based on actual experience, is that roofs should not be dependent on vapour barriers to avoid interstitial condensation associated with insulation. The most reliable system is certainly traditional 'cold roof' construction, in which the roof space is ventilated whether it is a pitched or a flat roof, the necessary thermal properties of the roof being achieved by providing insulation on top of the ceiling. In this connection it is useful to remember that foil-backed plasterboard functions as a heat reflector and improves insulation properties, as well as limiting humid air diffusion through ceilings.

13.4 Pitched roofs

It will be appreciated from the previous sections 13.2 and 13.3 that conventional pitched roofs perform reliably provided that there is adequate ventilation of the roof space and that there is adequate insulation at ceiling level, mainly through insulation provided between the joists but enhanced perhaps by the use of reflective foil-backed plasterboard. If these precautions are observed only problems with the roof covering are likely to occur.

Slates and plain tiles must be laid with double lap to prevent penetration through the joint between adjacent slates or tiles in a course. They are actually laid to a particular gauge or distance between courses or fixing battens. The gap between adjacent slates or tiles is protected by those in the courses above and beneath; the heads must extend upwards for a sufficient distance, or lap to provide adequate protection against wind-blown rain or snow, the required lap depending on the type and dimensions of the slate or tile, the pitch of a roof and the degree of exposure as explained in detail in British Standard BS 5534: *Slating and tiling* (this code of practice is issued in two parts, Part 1:1978 (1985) *Design* and Part 2:1986 *Design charts for fixing roof slating and tiling against wind uplift*). Part 2 draws attention to the fact that water penetration is not the only risk associated with slating and tiling and the resistance of the roof to wind is just as important. The great storm in October 1987 drew attention in many areas to the better wind resistance of slates compared with tiles, although this observation was actually very misleading; slates are usually more secure because each slate is always fixed with two nails, whereas plain tiles are often laid with only a single nail, and with some courses unnailed if the tiles are fitted with knibs which enable them to be hooked over the battens. In very exposed conditions it may be necessary to increase the lap simply to increase resistance to wind uplift; increasing the lap reduces the gauge and thus increases the overall frequency of nailing over the roof.

Pantiles are laid single lapped, that is the bottom of each tile projects for only a short lap over the top of the tile beneath and their design provides overlap protection at the side joints. Interlocking clay and concrete tiles are modified pantile designs which provide overlap protection of the side joints with often similar interlocking joints between the bottom of one tile and the head of the tile beneath. Obviously the lap is controlled by the design in interlocking tiles and it cannot be altered to allow for pitch or exposure conditions; instead the manufacturers define the lowest pitch at which their tiles can be used in relation to exposure. It is generally considered that slates can be used at lower pitches than plain tiles, and interlocking tiles can be used at even lower pitches, although problems often arise with interlocking tiles if the sections are damaged so that a prudent designer would not normally specify such materials at their minimum pitch. There is also a further problem in using interlocking tiles at low pitch, as they represent a rather high roof load and adequate support is more difficult to provide at low pitch; shallow pitch trusses distorted by interlocking tile loads are actually quite common, although only usually noticed if some trusses have additional support from partition walls so that the roof becomes distorted.

Sarking felt is now generally specified and required, but only as protection against wind-blown rain and snow. Sarking felt should therefore discharge into the gutter, but it is actually frequently trimmed short and water draining from the felt may then cause fungal decay development in rafter feet or gutter boards. Damage of this type can be particularly severe where there are other defects in the roof which mean that the sarking felt is more frequently subject to wetting, such as inadequate lap in relation to exposure or damaged interlocking tiles as previously described, although a much more common cause today is carelessness in tiling at the eaves. If sprockets are provided which reduce the pitch, it may be necessary to increase the lap at the eaves to prevent penetration, but even without such complications it is always necessary to provide a base course of short tiles or slates at the eaves. Water penetration problems into an eaves structure are often found to be due to omission of this base course, a defect that can often be missed immediately following construction if the sarking felt is effective, but which becomes apparent after some years when the sarking felt has deteriorated.

Although the theory of roof covering is relatively simple and straightforward, lack of understanding results in surprising difficulties. Hip valley gutters are usually formed using a relatively wide strip of lead lapped under the adjacent tiles, although it is sometimes preferred for aesthetic reasons to form a 'secret' gutter in which the tiles of the adjoining slopes are almost in contact, concealing a shallow lead box gutter. This arrangement is perfectly satisfactory in draining the roof, provided it does not become blocked with leaves. The usual problem is that leaves gain access to the gutter and form a blockage where it bends onto the sprockets, the gutter overflowing and typically causing staining of the ceilings and decay in the wall plates beneath. In one unusual roof the eaves terminated at the gables but the ridges projected well beyond them with raking fascia boards which were, in effect, functioning as both fascia and eaves boards in the sense that the bottoms of the tiles projected diagonally over them. It is impossible in this form of roof to provide a diagonal bedding course of tiles along these raking fascias and water penetration was therefore inevitable through the joints between adjacent tiles; the penetrating water accumulated on the sarking felt beneath, but it could not drain in the usual way but tended to lodge on the battens; dripping through the rear of the fascia and accumulating on the soffit boards beneath. Such designs are actually common in some countries, such as Japan, but they do not suffer from these water lodgement and seepage problems as soffits and sarking felt are omitted from the projecting roof section.

One problem with pitched roofs is the damage to lead lined valley and parapet gutters caused by organic acid discharges from tiles and stone roofs covered with moss and lichen growth; this problem is described in more detail in section 13.2. Another problem with hip valley gutters is that they may interrupt the insulation in an open roof without a ceiling so that the valleys may represent cold bridges and encourage interstitial condensation.

13.5 Flat roofs

The reliability of a flat roof depends on the reliability of the roof covering and the care with which it is applied, as previously discussed in section 13.2. However, the most serious problems are usually associated with interstitial condensation related to insulation, as discussed in section 13.3. Generally 'cold' roofs in which the roof space is ventilated and the main insulation is concentrated on the ceiling cause little problem, but 'warm' roofs which are dependent on a vapour barrier for protection against interstitial condensation are inherently unreliable, as it is difficult to construct a completely effective vapour barrier and perforation damage frequently occurs during or after construction.

Leaks through flat roofs are either random, or concentrated around a feature where the covering has not been properly formed or where differential movement is greatest, but sometimes water penetration damage is concentrated at the junctions between ceilings and walls beneath the edges of the flat roofs. In one Scottish castle, damage of this type developed beneath every parapet following roof repairs. The parapet copings were arranged to shed water towards the roof but they were trimmed flush with the parapet so that the water discharge was absorbed into the inner masonry face of each parapet. The problem was corrected by reconstructing the parapets to incorporate a normal damp-proof course linked with the roof asphalt upturn, but the interesting point is the reason for the sudden development of the problem; when the castle had been reconstructed in about 1908 the roof asphalt had been lead up the inner face of the parapet and across the copings to provide adequate protection against this form of penetration, but during repairs and alterations the architect decided that this asphalt protection was unsightly and inappropriate. Having specified the removal of the asphalt the architect unfortunately neglected to provide alternative protection in the form of a proper damp-proof course through the parapet!

Water penetration around the edge of a flat roof in another Scottish building was traced to blocked drains, the roof flooding and causing water percolation beneath the flashings over the asphalt upturns and downwards, to cause a band of dampness internally at the junction between the ceiling and the wall beneath. On one section of the roof a distinct band of dampness developed across the middle of a ceiling; the cause was the same as there was a steel joist across the roof supporting a wall above. One form of drainage from a flat roof is a hopper discharging at its base into a downpipe, the hopper often being partly concealed by the roof parapet and provided with a spout at a higher level to allow water to drain should the hopper become blocked. On one building it was observed that a hopper was blocked in this way and that the water discharge from the overflow spout was causing excessive wetting of the masonry beneath with dampness and erosion damage on the interior surfaces. It was arranged that the maintenance man would check and clear all the roof hoppers. A year later it was necessary to investigate complaints of continuing dampness but it was found that the hoppers were still completely blocked. The material blocking the hopper had decayed and was dense black in colour, and in wet conditions the surface of the dirt accumulation appeared to be an asphalt surface in the hopper; the maintenance man was cleaning only the overflow spout and did not appreciate that there was a hopper drain beneath which was completely obstructed by a deep accumulation of debris! This incident is sometimes described in lectures by the author on dampness or defects in buildings, and it is apparent from the response that many people consider that they would be unlikely to make such a fundamental mistake. In fact, flat roof hoppers blocked in this way, causing roof flooding or damage through overflow discharges, occur very frequently. The problem is very easily missed when inspecting a roof during rainfall when the wet debris in the hopper is similar in appearance to wet asphalt. Similar hopper blockage problems are also frequently found on lead roofs.

Felted roofs sometimes display a rectangular pattern of ridges, formed through blistering of the felt. One very large roof displaying this defect was inspected early in the morning during sunshine following rainfall, and it was observed that the areas between the ridges were still wet but the ridges themselves were dry and warm. The roof was steel construction with wood wool slabs supported by light steel sections. The slabs were designed to provide insulation, but the steel sections between them were conducting the heat from the roof space to the felt covering, and this heat was combining with the sun to soften the felt. The roof was over a swimming pool and the high humidity of the internal air provided a high water vapour pressure, sufficient to cause the softened felt to blister along the lines of the metal sections between each wood slab. Obviously the roof was also affected by serious condensation problems; condensation beneath the felt had generally saturated the wood wool slabs, but the junction sections at the edges of the slabs provided cold surfaces within the roof, causing heavy condensation which was dripping onto the ceiling beneath and which also caused corrosion of the lightly galvanised steel sections, this corrosion being encouraged by the swimming pool chlorination.

Floor problems

14

14.1 Introduction

A variety of floor problems have already been described in previous chapters. The purpose of this chapter is to draw attention to the problems that are most likely to be encountered and to describe some defects with unusual causes.

14.2 Solid floors

Movement is certainly the most serious problem in solid floors, as explained in sections 2.3, 3.5 and 4.2. Although movement joints must be provided to accommodate shrinkage and expansion arising from fluctuations in temperature and moisture content, it is the changes during and following construction that actually cause the most serious problems. In a building with a reinforced concrete frame, the reinforced concrete upper floor slabs are integral with the frame and the whole structure suffers uniform movement due to fluctuations in temperature and moisture content, particularly shrinkage following construction. If rigid ceramic tiles are laid on the floors, it is essential to limit the size of the tile bays and to provide movement joints between the bays, as well as between the tiles and the walls. If these movement joints are omitted the shrinkage of the concrete floor slab will tend to dislodge the rigid tile flooring, but in addition the shrinkage of the structure will compress the tile flooring and doming may develop. Sometimes the tiling becomes detached from the floor through a shock wave related to some incident, such as something dropped on the floor, flexing of the building through wind or earthquakes, or even an explosion in the vicinity, the tiles then being thrown upwards suddenly, perhaps injuring persons within the room; sudden detachment of tiles can also occur from walls, as described in section 11.3, and the same problems can occur with rigid stone flooring such as marble or granite slabs.

Solid ground floor slabs are sometimes integral with a reinforced concrete structure, as for upper floors, but they are often separately supported on the ground beneath and isolated from the structural concrete and walls by movement joints around their perimeter. Ground floor slabs do not represent any special problems compared with upper floor slabs, except that the perimeter movement joints should coincide with movement joints in applied screeds and tiling.

Concrete floor slabs are most usually cast with a finished surface but are topped with a separate screed finish; the mix is basically either a fine aggregate concrete or a mortar, as described in more detail in sections 7.4, 8.2 and 8.4. Care must be taken to bond the screed satisfactorily to the slab, preferably by laying the screed as soon as possible after the slab is set or, if a fully hardened slab is involved, by brushing off all loose dirt and by wetting the surface before laying the screed. Risk of loss of bond to the base slab can be reduced by ensuring that the screed is not too strong and not too thick; mix proportions of cement to fine aggregate of 1:4 by mass are usually recommended with screed thickness not exceeding 40 mm, although bonded screeds should not be less than about 25 mm thick. If a screed thickness in excess of 50 mm is necessary, a fine concrete comprising 1:4.5 cement to total aggregate should be used to minimise shrinkage and match the screed properties closely with those of the supporting concrete; large aggregate grading should be limited to 10 mm in fine concrete for these screeds.

One of the most common problems in screeds is random fracturing or crazing due to excessive shrinkage of the screed; if the bond to the concrete is also poor, curling or lipping at bay edges will also develop. Shrinkage may be due to the use of an excessively cement rich mix, although it is more often due to an excessive water to cement ratio; this may be caused by carelessness in mixing, but usually by the use of very fine aggregate which makes it necessary to add excessive water to maintain workability, a problem that has been described in section 7.4 in relation to render. Many floor layers have encountered crazing and curling problems in this way and prefer to use relatively dry mixes; in fact, such mixes result in inadequate distribution of the cement, particularly if normal drum mixers are used, and such mixes are also difficult to compact, resulting in a screed with a rich surface where it has been worked but a weak base, so that the screed can be easily damaged by stiletto heels and similar high loads.

Screeds should be provided with movement joints coinciding with those in the supporting concrete. When a screed is followed by a rigid finish, such as ceramic tiles or stone slabs, the movement joints in the floor topping should coincide with any movement joints in the screed and supporting concrete, with additional movement joints to divide the flooring into appropriate size bays.

Floating screeds, in which screeds are isolated from the base concrete to reduce sound transmission, present special problems. They are particularly sensitive to curling through excessive shrinkage as they are not bonded to a rigid substrate, and they must be sufficiently thick and strong to withstand normal loads; breakdowns in such screeds are generally due to the use of thicknesses of less than 50 mm, or excessive friability due to the use of excessively dry mixes.

Polymer modified cement is sometimes used to form a thin screed or as a separate topping to form a so-called 'seamless' floor. In fact such toppings can only be seamless if applied to concrete floors that are integral with their supporting structure, so that they do not contain any movement joints; in all other cases movement joints are necessary in the topping and must coincide with those in the supporting floor, and fractures are inevitable if these movement joints are omitted, a very common failure with proprietary flooring systems. Resin and aggregate systems can be similarly used where improved chemical resistance is necessary; polyester, epoxy and polyurethane resins are preferred, applied as resin screeds or thin coatings. Polymer modified cement systems are usually based on acrylic or similar emulsions and can give excellent and inexpensive general purpose finishes, but they will be damaged by the same chemicals that affect cement, including sugar syrup solutions in food and drink works which can cause serious expansion problems. Floors of this type are sometimes offered with a polyurethane finish to give protection against spillage, but it must be appreciated that the polyurethane is only effective provided it remains intact, and a thin floor coating of this type is particularly susceptible to abrasion damage. Chemical resistant floors based on mixtures of resins and aggregates can be fully-filled impermeable systems which only need a final finish for aesthetic reasons, or because a particularly smooth surface is required for ease of cleaning or decontamination. However, some systems contain less resin, or resin diluted with solvent for ease of application, and produce permeable systems which must be finished subsequently with either the original resin or a polyurethane system if complete chemical resistance is required, as otherwise any spillages may seep through and damage the supporting concrete. Polyester, epoxy and polyurethane resins will only polymerise or 'cure' if the temperature is sufficiently high. The most common cause of softness is low temperature, usually because the temperature of the air above the floor is adequate but the substrate on which the resin is being laid is too cool.

Ceramic tiles are often used where chemical resistance is required, particularly in bottling works. However, whilst the tiles may have excellent chemical resistance the flooring system will only be as good as the grout surrounding the tiles and the bedding on which they are placed. If ordinary cement systems are unsuitable because of chemical or sugar spillages, epoxy systems are preferred.

Wood is used extensively as a solid floor surface, both as a heavy traffic surface where tar impregnated blocks are usually used, or as a decorative strip or block floor. In normal decorative floors the finishes are critical, but there are usually no special moisture problems and modern synthetic resin systems, particularly polyurethanes, can give very attractive finishes with excellent wear resistance; a matt finish is preferable to gloss as abrasive wear is less obvious. The main problems with timber floors are related to the generally high cross-grain movement of wood. These movement problems can be minimised in theory by kiln drying flooring wood to the average moisture content that it will achieve in service, typically about 8% for a centrally heated building. However, if the wood is delivered to site and laid in this condition it will tend to absorb water from the atmosphere, which will be excessively humid because of other damp parts of the structure, and the wood will expand, in extreme cases moving internal partition walls or doming. If the wood is, alternatively, conditioned on site before laying it will have a much higher moisture content and the subsequent conditioning of the building will result in drying and excessive shrinkage. It will be apparent from these comments that shrinkage problems in wood floors cannot be avoided simply by kiln drying to an appropriate moisture content; the only realistic solution to this problem is to use only woods with low movements.

14.3 Suspended floors

This section is concerned only with suspended wood floors; suspended concrete slab floors are considered, for convenience, in section 14.2.

Suspended floors generally cause fewer problems than solid floors. A traditional suspended floor comprises boards supported by joists. The main problems are concerned with the cross-grain movement of the boards, as described for wood block and strip floor coverings on solid floors in section 14.2. It is unrealistic to use only wood with low movement for normal boarded floors; instead it has become normal practice to use tongued-and-grooved boarding to prevent actual gaps developing and to avoid draught problems when the floors are spanning a ventilated sub-floor space. However, the gaps between the boards, as well as doming and cupping of the boards themselves, are very unsatisfactory, and a better solution is the use of chipboard which is much less affected by movement. It is important to use only a grade of chipboard which is designed for use as flooring to ensure adequate strength; a few years ago it was quite common for chipboard floors to fail locally, particularly under bed legs which represent the highest floor loadings in domestic situations. Chipboard is not ideal for this purpose as it will deflect progressively or creep under continuous loads, particularly if it is subject to fluctuations in moisture content, as in suspended ground floor situations, but these problems can be avoided by using a suitable grade of plywood. Chipboard and plywood manufactured for flooring purposes are usually provided with tongued-and-grooved edges; it is important when using chipboard or plywood flooring to ensure accurate joist separation so that joints coincide with joist supports, and also to lay sheets the correct way up, most flooring sheets being marked on their underside with a warning.

The most serious problem with suspended floors is the damage caused by electricians and particularly plumbers, as described in section 2.2. Damage through notching is not confined to weakening of the joists, but also affects the floor boarding. This is a problem that arises largely through poor design and management; it is urgently necessary to give more consideration to pipe and wiring runs, as well as to ensure that any weakened joists are repaired using joist straps, and piping and wiring are completed before floor boarding is laid with carpenters, rather than electricians or plumbers, providing proper access traps.

Wood board, chipboard or plywood floors are often formed on battens laid on concrete slab floors, particularly in modern buildings in which extensive wiring services are required as the floor cavities can be used to conceal the wiring, whilst still allowing access for alterations if suitable traps are provided. Normal sawn battens or shallow joists can be used if they are firmly fixed to the concrete substrate, but this firm fixing results in sound transmission from the floor to the concrete structure which may be unacceptable. This problem can be avoided by providing a suitable sound isolating support for the battens. Such floors are often designed with the battens resting on glass- fibre quilt; this system is not very effective as the quilt compresses under load, losing its sound isolating properties, and also giving an unstable floor. A more efficient proprietary system uses foam plastic applied to the bottom of the battens, but problems occur with both these systems through irregularities in the battens and the concrete, which may result in irritating instability in the floor. The best proprietary systems involve supports, sometimes known as acoustic clips, which are fixed to the battens and to the floor slab. With all these systems it is also essential to isolate the edges of the floor platform from contact with the walls.

The risk of fungal decay development with such floor systems is seldom appreciated. If these batten floors are laid over a ground floor slab or particularly over a suspended slab over an open space, the floor space is much cooler than the accommodation above the floor and there is a danger of condensation within the floor space. For example, if the accommodation air has a relative humidity of 70% at a temperature of 20 °C, it has a dew point of about 14.5 °C; this dew point can be derived from Figure 4.1 in section 4.2. Concrete slabs supported on the ground are generally cooler than this, unless the slabs incorporate substantial insulation, and condensation will therefore occur on the slab surface and may also occur within the air space and affect the battens. It is therefore essential that the flooring and the supporting battens should be adequately preserved. The danger is, of course, particularly acute during construction when construction moisture will contribute to the humidity and increase the condensation risk. A special danger also arises when there is a fibreboard movement joint beneath a floor of this type, as this condensation risk ensures that fungal decay will develop in the fibreboard and may spread to the floor structure under suitable conditions; it was found in a large public authority building in Ireland that almost all the fibreboard movement joints had decayed in this way, although a specially treated fibreboard had been used which was designed for this purpose and in many cases the decay had spread into the floor structure, causing collapse of the floor in one area even before the building was occupied.

Combustion problems

15

15.1 Introduction

Heat energy in buildings is often provided by combustion of solid, oil or gas fuels. This chapter is not concerned with the detailed design, construction and operation of combustion equipment, but only with some of the problems that may be encountered with normal types of equipment. Installations in large commercial and industrial buildings are usually designed and installed by experienced specialists and give fewer problems than normal domestic systems. Although architects are interested in the appearance of fittings such as radiators and fireplaces, they seldom have any knowledge of combustion technology, relying on the general contractor to supply and install suitable fireplaces, and the plumber and heating engineer to provide suitable boilers. Appropriately qualified and experienced plumbing and heating engineers can provide an efficient and reliable installation, but the incidence of defects suggest that many firms lack qualifications and experience, or even common sense in some instances.

15.2 Combustion air

Efficient combustion is only possible with an adequate supply of fresh air. In traditional construction the combustion air for an open fire is derived from diffusion through windows, doors and suspended floors, but the flow of combustion air in this way also causes unpleasant draughts. The more extensive use of solid floors, and particularly improvements in the sealing of windows and doors to save heat energy and improve comfort, have resulted in virtually sealed rooms in which there is no obvious source of combustion air. In such situations an alternative source of combustion air is essential. If the chimney is on an outside wall the combustion air can be introduced direct from the exterior to beneath the hearth, but where the fireplace is on a partition wall it will be necessary to provide a duct into a ventilated space beneath a suspended floor or direct to the exterior with a solid floor.

Although there is increasing use of boilers provided with direct combustion air and flue gas ducts through an exterior wall, with combustion completely isolated from the interior air, conventional free standing and wall mounted boilers fed by the accommodation air are still extensively used, their installation instructions always emphasising the need for a low-level vent through the exterior wall to ensure adequate availability of combustion air. The draught induced in a tall boiler chimney or caused by a blower in some forms of boiler can be substantial; if the air vent is omitted boiler firing will reduce the pressure in the accommodation as it attempts to draw in combustion air. If the internal doors are open or not particularly well sealed this reduction in pressure will affect most of the accommodation and induce a powerful down draught in any open flues. If open fires smoke badly this cause can usually be diagnosed by observing that the smoking is worst when the boiler is firing. The problem can only be entirely corrected by providing both a combustion air vent for the boiler and separate air inlets for the open fires.

Gas cookers and portable bottled gas heaters do not produce harmful combustion products; if they are working efficiently, they produce only water and carbon dioxide, the products of normal breathing. However, if they are used in the absence of adequate ventilation they will progressively reduce the oxygen concentration and increase the carbon dioxide concentration, causing a feeling of stuffiness and shortness of breath, usually accompanied by excessive condensation on the windows. If ventilation is still not provided the inadequate oxygen supply will eventually result in the formation of poisonous carbon monoxide. Adequate combustion air through vents from the exterior is therefore essential. In a kitchen the vent provided for an open flue boiler is sufficient, but if a sealed flue boiler fitted to an external wall is used, or a boiler is installed elsewhere, a special vent must be fitted in the kitchen to provide combustion air for gas cooking, a requirement that is often overlooked and which explains the stuffiness in some kitchens when the cooker is in use for prolonged periods. A vent for the products of combustion is also necessary, at least a high-level wall vent but preferably an extractor over the cooker, which will also remove cooking smells and prevent them from diffusing into other rooms; an extractor can be induced draught, that is a simple flue system, although modern systems are generally fan induced and operated only during cooking. It is important to appreciate that some cooker hood systems are not extractors but simply operate as odour removers by circulating the air through filters and activated charcoal.

Hot water is sometimes supplied by instant heating systems in which a gas flame heats the water as it flows through a pipe. Such heaters were formerly extensively used for kitchen sink and bathroom hot water supplies, although bathroom heaters caused a number of fatalities which have discouraged their use. In cold weather the bathroom hot water heaters were often operated for a prolonged period with the bathroom door closed, limiting the combustion air supply. With a tightly sealed door a situation would eventually develop in which reduced pressure in the room would prevent draught in the flue, and the flue gas would then enter the bathroom and recirculate. The oxygen concentration in the air would progressively decrease and the carbon dioxide concentration increase, until eventually poisonous carbon monoxide would be formed. Fortunately the dangers have been appreciated and most hot water heaters of this type have been removed from bathrooms, although some are still operating in kitchens, and similar dangers can occur with portable bottled gas heaters, as previously explained.

A more common problem in domestic situations is progressive obstruction of the air inlets to the burners on gas fires and boilers, restricting the air supply and resulting in a yellow and eventually smokey flame. Obstruction in this way occurs particularly in homes with dogs and cats, as it is their fur that is drawn into the air inlets! Routine servicing of burners should include cleaning the air inlets, although this is sometimes overlooked, a heating engineer adjusting the mixture in an attempt to obtain more efficient combustion whereas a yellow flame is more often caused by simple obstruction of the air supply.

15.3 Flue defects and deterioration

The function of a flue is to convey the products of combustion to the exterior. The main products are water and carbon dioxide, but oil burning produces sulphur dioxide and solid fuels produce nitrous oxide in addition. In a normal open fire there is a generous flow of air which is usually sufficient to enable the combustion products to leave the chimney before condensation of the water content, at least once the flue has reached normal operating temperature. However, if the air flow in the flue is restricted, either because there is inadequate combustion air or because the flue is fitted with a damper, the flue gas and lining will be cooler and the water content perhaps higher, increasing the danger of condensation which will then absorb sulphur dioxide and nitrous oxide to form sulphurous and nitrous acids, although the presence of bacteria will convert these to more corrosive sulphuric and nitric acids as explained in more detail in section 7.6. The mortar joints and any mortar lining will be affected by these acids, typically producing hygroscopic salts which account for the persistent bands of dampness on internal chimney breasts and external gables which identify the positions of flues.

Solid fuel boilers are generally controlled by restricting the air flow, older boilers also being fitted with flue dampers, although obviously the dampers must not be closed when the air flow is open. Usually solid fuel boilers are coupled direct to flues and very considerable condensation and hygroscopic salt problems can be caused. It will be appreciated that damage is not confined to the direct corrosive action of the flue acids on the mortar, but sulphate attack will also occur, causing disruptive expansion of the mortar followed eventually by loss of cohesion, as explained more fully in section 4.4; for this reason sulphate resisting cement should always be used in preference to ordinary Portland cement for constructing flues. Alternatively, damage to the mortar can be avoided by using a stainless steel liner, but with slow moving flue gas condensation is inevitable, generally dripping back into the fire in the case of a solid fuel boiler. However, similar condensation can occur with oil and gas boilers. Some lining systems are provided with a reservoir at the bottom to collect the condensation, the boiler venting into the side of the flue pipe, but it is much better to allow a free flow of air into the base of the flue as this avoids condensation by diluting the combustion gas and is, in fact, the most efficient way to avoid flue problems in both conventional and stainless steel lined flues. Whilst gas boilers often use the configuration that has been described, that is an open bottom to the flue with the combustion gas being introduced into the side of the flue higher up, some oil boilers use a flue directly connected to the fire box but with a balanced door in the side of the flue to permit free air entry; the weighting of the air inlet door is critical on this type of boiler and must be carefully set to protect the flame from wind effects in gusty conditions, whereas the open ended flue with the combustion gas entering at the side gives complete protection against wind effects.

Care must be taken to ensure that flue gases are unable to enter the accommodation. Problems are best avoided by using cast concrete interlocking flue liners rather than relying on normal brickwork or blockwork masonry. The type of problems that can occur are best illustrated by an example which concerned a new house in the Isle of Man. A young girl developed breathing difficulties which were worst in the early morning during the winter months, when it was also observed that the air in her bedroom had a distinctly acrid smell, similar to the smell of burning coal. Problems were particularly severe when an open fire had been used in the room beneath, and investigations showed that combustion products from the fire were entering the wall cavity and then seeping into the bedroom through gaps in the inner leaf around the floor joists bearings. The main problem was found to be a lack of care in forming the throating connecting the fireplace to the flue, but further leakages occurred from the flue itself, particularly at a bend at first floor height.

With gas and oil boilers it is particularly important to ensure that the fuel and air mixture is properly adjusted. Inadequate air will result in sooting, but excessive air will result in an unstable flame and corrosion where the flame impinges on water jacket sections or baffles.

16

Plumbing problems

16.1 Introduction

This chapter is not intended to be an appraisal of plumbing technology, but only a brief account of some of the problems that may be encountered in defects investigations in buildings.

16.2 Hard and soft water

A soft water lathers easily with normal soap, but it is much more difficult to obtain a lather with hard water. Hardness is caused in two different ways. Permanent hardness is due to the presence of sulphates, chlorides and nitrates, usually as salts of calcium and magnesium, although iron and sodium may also be present. This type of hardness is described as 'permanent', as it cannot be removed by boiling the water. Temporary hardness can be removed by boiling and is due to the presence of bicarbonates, but loss of carbon dioxide results in conversion to carbonate and the familiar scale of hard water areas through formation of calcium carbonate. Some water supplies are treated with lime which reduces particularly temporary hardness by reducing the calcium and magnesium bicarbonate contents. Domestic water softening usually involves ion exchange resins which remove troublesome calcium and magnesium ions from water by absorption; the resins eventually become saturated and need to be reactivated, usually by passing common salt solution through the resin, releasing the absorbed ions and allowing them to be discharged to waste. Dishwashers often incorporate water softeners of this type with automatic reactification; it is important to maintain the level of softening salt, as well as the main washing detergent and the drying aid. Some water softening systems, particularly those used in laboratories to prepare deionised water, use cartridges which are exchanged for reactification.

The formation of hard water scale is certainly the most serious routine plumbing problem and the justification for lime softening of hard water supplies, although lime softening is not usually completely effective in removing bicarbonate and some scale still forms but at a much slower rate. Scale formation is most noticeable in kettles in which boiling of each filling of fresh water contributes a little more scale. The same problem occurs when water is fed directly through a boiler, modern boiler systems avoiding this problem by heating the hot water supply indirectly; the boiler circuit is isolated and only subject to insignificant scale through the initial water fill, although some scale can still develop on the heating coil in the calorifer or cylinder. Such scale formation problems are much less severe and much less dangerous than with direct heating of the water supply within the boiler, which may involve blockage of the boiler and a danger of explosion. Radiator systems are, like indirect hot water heating systems, isolated from the water supply and only suffer from insignificant scale formation when the system is filled with fresh water. Scale formation is not confined to kettles or boilers, but also occurs wherever water evaporates, forming a water level mark in WC pans and cisterns and also obstructing taps and ballcock valves.

16.3 Metal corrosion

Metal corrosion in plumbing is most severe with acid soft water which can cause direct corrosion of lead and copper pipes, sometimes resulting in excessive lead or copper content in the water. Lead is the most serious problem and the use of lead supply pipes is now discouraged or forbidden by most water companies. The most severe corrosion problems result from electrochemical action through the use of different metals connected together in a system. If iron and copper pipes are connected, progressive corrosion of the iron will occur as explained more fully in section 4.5, this corrosion often accounting for leaks in supply pipes or where copper pipes are connected through control valves to steel radiators. Corrosion inhibitors can be added to closed circuits such as radiator systems and calorifier circuits in hot water systems and can very effectively reduce corrosion, although some proprietary systems are distinctly more efficient than others. These variations in properties may be associated with the ability of the corrosion prevention chemicals to encourage or prevent development of sulphate reducing and similar bacteria that can develop in these anaerobic closed circuits, despite their high temperatures; anaerobic bacterial problems are described in more detail in section 4.5.

16.4 Drains

Routine problems with drains are generally associated with obstructions which can often be attributed to unsuitable falls. If the falls are inadequate or vary, foul water drains will be unable to clear solid materials, but if the falls are too steep the solids may become stranded through excessively rapid drainage, again resulting in blockage. Falls are therefore critical, but foul drain runs must also be constructed to avoid sumping through settlement. Blockages usually occur at bends and it is therefore prudent to provide a manhole or rodding eye at every bend to facilitate rodding.

Water and sediment in foul water drains rapidly becomes anaerobic through the high organic content of the sewage, encouraging the development of anaerobic bacteria, such as sulphate reducing bacteria. These bacteria encourage rapid metal corrosion of iron and steel pipes, causing premature failure in sewers with intermittent flows, as described in section 4.5; where there are continuous flows the sewerage is often sufficiently aerobic to avoid these problems. Sulphate reducing bacteria liberate hydrogen sulphide gas into the air space above the sewage and this is absorbed by condensation where it is rapidly converted to sulphate by the action of oxidising bacteria, causing severe sulphate attack damage to mortar joints, rendered surfaces and concrete as described more fully in section 8.6. Sulphate resisting cement should always be used in preference to ordinary Portland cement in conjunction with foul drains and sewers.

Unplasticised polyvinyl chloride or uPVC is now extensively used for waste and overflow pipes. One of the features of this material is its high thermal movement which can result in excessive stressing of firmly fixed pipes, as well as noise problems through movement of pipes carrying alternately hot and cold water. Distortion and fracturing of firmly fixed pipe runs can occur, although movement problems are seen most frequently in push fit cistern overflow pipes which are often inadequately supported, simple diurnal and seasonal temperature changes being sufficient to cause severe sagging and disconnection of the joints with serious consequences should the cistern overflow and discharge into the pipe.

External problems

17.1 Introduction

This chapter is not intended as a comprehensive review of the many external problems that affect buildings, but only draws attention to certain problems of particular interest or importance. Ground levels in relation to internal floor levels are obviously very important. In normal free-standing construction all horizontal damp-proof courses giving protection against rising dampness should be at least 150 mm (6") above adjacent external ground level. On a sloping site this is best achieved by excavating into the rising slope to provide a protective clear area around the building, but this may not be possible on a steeply sloping site, and it may be necessary instead to adopt tanking or, in effect, basement construction. Such problems have already been considered in detail in Chapter 4, particularly in sections 4.8, 4.9 and 4.11.

17.2 Trees

Trees close to buildings always present a risk of damage. The obvious risk is damage due to branches breaking or the entire tree falling in windy conditions. Certain soil conditions, such as a thin depth of soil covering rock or clay, will encourage surface rooting and a danger of the tree falling, particularly when it becomes over-mature and the root system begins to deteriorate as the tree reaches its maximum height. Over-maturity is also associated with a decline in vigour which makes branches more brittle, broken branch stumps then allowing fungal infections to develop leading to heart rot and progressive deterioration. Obviously some species of trees are much more susceptible to damage and deterioration than others, but care must be taken to ensure that trees do not become over-mature in situations where they may damage buildings. Trees close to buildings are also the main cause of blocked gutters and fall pipes, as discussed in more detail in sections 13.4 and 13.5.

Damage caused by tree roots to foundations is an extremely serious problem; it has already been discussed in detail in section 10.3.

17.3 Paving and tarmacadam

The usual problems with external paving and tarmacadam are uneven settlement. With patio and path areas designed only for foot traffic, firm compaction of the ground is theoretically all that is required to provide adequate support, but deep fill will progressively compact naturally even in the absence of imposed load and may cause serious problems. Suitable precautions are normally adopted on sloping or landfill sites to provide adequate support for a building, such as deep strip, pile or raft foundations, but similar precautions are seldom adopted for ancillary features. As a result access steps and paths, and even porches and bay windows, sometimes suffer from serious settlement. Swimming pools are often constructed on sloping sites to minimise excavation by utilising the slope to provide the shallow and deep ends, but the excavated soil is often used around the pool construction with inadequate compaction and pool surrounds often suffer serious settlement as a result.

The deterioration of tarmacadam surfaces may be due to the loss of an excessively volatile binder, the cause of the brittle fracture problem in asphalt roofs described in section 13.2. An insufficient binder content in relation to the aggregate may also result in a weak and loose surface; the binder requirement can be minimised by carefully grading the aggregate to provide maximum aggregate fill, as described for concrete in section 8.4. However, an inadequate depth of tarmacadam is the usual explanation for fractures and weakness. Plant growth will also cause disruption of tarmacadam and is particularly encouraged if the depth is inadequate and the finish rather permeable due to the use of insufficient binder; plant growth can be discouraged by treating the ground thoroughly with weed killer before laying tarmacadam, and also by the use of coal tar binders rather than bitumen derived from petroleum.

Algal, moss and lichen growth can be troublesome, particularly on cast concrete and concrete slab surfaces. Suitable biocides have been described in section 7.7.

17.4 Fencing

Boundaries between adjacent sites or gardens may be provided by brickwork or stone masonry walls, which have already been considered in detail in sections 7.2 and 7.3, or by fences constructed from various materials. It is not generally appreciated that the most durable fencing is actually properly preserved wood. Copper-chromium-arsenic water-borne salts to BS 4072 or creosote to BS 144 are completely reliable if treatment achieves adequate retention and penetration, usually involving vacuum-pressure impregnation of wood species in which sufficient penetration can be achieved; European redwood or Scots pine is the most suitable softwood species available in the British Isles, as explained in section 6.7. Rails and boarding which are not in ground contact are not at such risk and a lower level of preservation treatment will be adequate; for example, the same type of treatment applied to European whitewood or spruce which is resistant to penetration will be adequate for such purposes. Generally round redwood poles are more durable than square poles, because the round poles possess a complete envelope of easily impregnated sapwood whereas sawn poles expose heartwood which is more resistant to treatment, although it possesses some natural durability. The end-grain is much more easily impregnated in both heartwood and sapwood, although it is important for posts to be treated in their final length as cutting may expose untreated heartwood; decay in properly preserved fencing wood is always associated with untreated wood exposed by cross-cutting.

Wood posts are often used to support plain wire, barbed wire and wire mesh fencing. The life of the wire depends directly on the thickness of the protective zinc galvanising, as the zinc coating is progressively lost until the steel is exposed and corrosion commences. Corrosion is most rapid where the galvanising is abraded by fixings, particularly in polluted urban atmospheres and in fences subject to sea spray or road de-icing salt. Metal posts suffer similar damage, although initial failures are generally concerned with abrasion where wires pass through post holes or fixings. Metal post life can be considerably extended by a bitumen or particularly a coal tar coating; unfortunately post coatings are usually applied only to the exposed post after erection, giving no protection at the ground line where corrosion is most severe.

Concrete posts have a strictly limited life, deteriorating through corrosion of the reinforcement which causes fracturing of the surface concrete. Corrosion occurs only after carbonation of the concrete has reached the depth of the reinforcement, the small dimensions of fence posts limiting the amount of concrete cover which can be provided over the reinforcement and thus limiting the life of the posts. The life of concrete fencing panels is even less because of their limited thickness. Concrete failure is discussed in more detail in Chapter 8.

18

Health and safety problems

18.1 Introduction

There are many health and safety problems associated with buildings. Indeed, most accidents occur in the home, although this is not an indication of extreme danger associated with the house and its content, but simply a reflection of the amount of time that we spend in our homes. It is not intended that this chapter should be a comprehensive review of health and safety problems associated with buildings, but that it should instead draw attention to particular problems and discuss those which currently create most public concern.

18.2 Ventilation

Inadequate ventilation is one of the most serious problems with modern buildings, arising mainly through fundamental changes in living habits over recent years, but exaggerated by an unreasonable preoccupation with thermal insulation as a means to achieve energy conservation and improved comfort.

Ventilation, which is normally measured in air changes per hour, is necessary to ensure an adequate supply of oxygen for breathing and combustion, and to remove the products of our living processes, that is carbon dioxide and moisture from breathing and cooking, as well as additional moisture from bathing and laundering. Ventilation problems were unknown before World War I, as it was normal for all living spaces, including bedrooms, to be provided with fireplaces and flues inducing adequate ventilation even without a fire; indeed, a flue would normally induce excessive ventilation and fireplaces were fitted with dampers to seal the throat when the fire was not in use, the most common type of dampers comprising a tilting flap at the top of the fire back or a damper projecting from the back of a hood which could be pushed in to close the flue. There was also sufficient air flow around a closed damper to induce adequate ventilation in a room through the window frames or door. Following World War I, the style of fireplaces progressively changed and dampers were abandoned, leaving open throats which induced excessive draught, and this change and the need for reduced costs resulted in the construction of houses without fireplaces and flues in the bedrooms. The limited ventilation encouraged condensation problems and induced a feeling of stuffiness, and some local authorities introduced building by-law requirements for wall ventilators, designed so that they could not be closed, for bedrooms without flues.

The progressive introduction of full central heating after World War II further reduced the dependence on open fires with the introduction of automatic oil and particularly gas fired systems which could be relied on to maintain adequate temperatures, and eventually open fires were completely omitted from many new houses. Central heating was only realistic because of the relative low cost of oil and gas fuel, but the rapidly escalating demand for fuel strained resources and threatened long-term reserves, creating the petroleum crisis and an increasing awareness of the need for energy conservation whilst maintaining the improved comfort to which people had become accustomed. Reductions in structural heat loss were achieved relatively easily, but the need for reductions in ventilation heat loss has not been approached in such a realistic way, any savings resulting from the relatively unscientific wish to reduce draughts.

As a result modern houses are carefully sealed and there is virtually no ventilation. A few years ago useful continuous ventilation was provided by a boiler installed in the kitchen, but it became fashionable to have the boiler in a laundry, garage or even a separate boiler room or compartment, and modern compact boilers have air inlets and flues ducted directly to the exterior with no connection with the interior air, all these changes reducing ventilation. As a result, the air in modern homes in the winter suffers from a low oxygen content, high carbon dioxide content and high humidity which are apparent as a feeling of stuffiness, with excessive condensation even in well insulated structures. Although open fireplaces in living rooms are again becoming fashionable, difficulties are often encountered in maintaining adequate draught because of the sealing of the rest of the accommodation and the lack of adequate combustion air.

These various problems have been discussed previously in sections 3.2, 3.3, 4.7 and 15.2. The correct solution is, of course, controlled ventilation to ensure an adequate but not excessive air change. One method which is particularly suitable for a flueless house is a small fan feeding air from the roof space into the accommodation through the ceiling of the landing or, in a bungalow, the hall, an arrangement that slightly pressurises the internal accommodation and achieves ventilation without draughts and without unnecessary heat loss. Another alternative is a passive system in which ventilators are provided which are fitted with a special membrane constructed from a material which will allow air diffusion as well as condensation dispersal; the membrane is cold so that condensation is encouraged, but when condensation occurs on the membrane it is absorbed and dispersed to the exterior by evaporation.

18.3 Accidental fire dangers

Accidental fire in buildings is almost always initiated in the contents rather than the structure, the most common cause being ignition of furnishings through an electrical fault, lighted fuel falling from a fire or a dropped cigarette. Structural fire precautions are necessary to protect exit routes and to reduce fire spread; masonry materials, concrete and gypsum plaster are particularly useful as they are nonflammable, although wood is an excellent fire barrier and solid wood will perform particularly well in fire doors because of its low thermal conductivity and slow fire penetration properties.

Fire precautions will not be considered in detail, although it is necessary to draw attention to certain problems that are sometimes encountered. It is particularly important to prevent a fire originating in the day accommodation from spreading to the bedrooms, and fire in one house from spreading to another in semi-detached and terraced properties. In conventional accommodation these requirements are usually achieved by the use of plasterboard ceilings and fire stop gables in roof spaces constructed of brickwork, blockwork or timber frame covered with plasterboard. However, it is not generally appreciated that it is essential for this fire barrier to be absolutely intact without any gaps which may allow fire penetration. It is not sufficient for the fire stop to be fitted tightly under the roof sarking felt; a strip of mortar should be laid between the battens on top of the felt to continue the barrier up to the roof tiles. If a plasterboard barrier is used, the joints must be covered with strips of plasterboard to prevent penetration.

As accidental fire originates in the contents rather than the structure, a wood frame building does not involve any greater risk of fire than a conventional masonry building. However, if fire in a wood frame building is able to penetrate through the inner walls into the cavity beneath the cladding it can easily spread upwards to affect the structure above or sideways to affect adjacent units, and it is therefore essential that wall cavities should incorporate horizontal fire stops at each floor level, as well as vertical fire stops to prevent spread between units or into party wall cavities. Obviously, fire stops must completely close the cavities to be effective, although it will be appreciated from the comments in section 11.4 that the ties used to restrain cavity width in brickwork clad timber frame construction have been largely ineffective in the past and cavities are often found which have widened, destroying the effectiveness of the fire stops. Fire stops should therefore be constructed from a resilient material, such as a fire retardant foam which will be able to tolerate some changes in the cavity width; fire stops should also be able to tolerate any rainwater penetration of the external cladding, particularly when brickwork cladding is used, so that wood used in fire stops should be preservative treated.

Remedial wood preservation treatment of buildings often involves the use of flammable solvent preservatives, as described in section 6.7. There is obviously a risk of accidental fire whilst spraying these preservatives and subsequently whilst the solvent evaporates from the treated wood. This risk can be substantially reduced by ventilation, but ignition danger should also be minimised, not only by prohibiting smoking but by isolating wiring in the treated area and by using only flame-proof lamps and equipment. These standard precautions are observed by all prudent and competent remedial timber treatment contractors; they are, in fact, part of the code of practice of the remedial treatment section of the British Wood Preserving and Damp-proofing Association. However, accidental fires are still caused by remedial wood preservation treatment, even when these precautions are observed. One possible cause of fire is electrostatic discharge, cause by the flow of solvent in the hose and its discharge through the spray nozzle. This cause of ignition is very rare; ignition through the cold spray contacting hot ordinary light bulbs is by far the most common cause of fire through operatives using installed lighting or unsuitable lead lamps. Another frequent cause of fires is auto-ignition; if a fibrous material such as glass fibre insulation quilt is saturated with solvent, the enormous surface area of solvent in contact with oxygen in the air will result in slow oxidation, which will be sufficient to increase the temperature of the interior of the quilt progressively until ignition eventually occurs. Obviously insulation materials must always be removed before applying organic solvent preservatives, partly to avoid this danger but also to expose the surfaces of the timbers so that they can be properly treated.

Remedial treatments with organic solvent systems can also affect electric wiring, dissolving the plasticisers in PVC insulation. The insulation becomes brittle in extreme cases and may flake off the wire, introducing a danger of shorting and fire. In addition, the dissolved plasticiser may drip onto plaster ceilings or spread into surrounding plaster where wires are concealed in walls, causing unsightly staining which is difficult to conceal with paint or wallpaper.

18.4 Methane dangers

Methane CH_4 gas, the 'fire-damp' that is so feared as a cause of explosions in coal mines, is emitted naturally from all soils containing organic material. Methane emissions from rock fissures above oil deposits were the cause of the eternal fires of Baku and today similarly formed methane is collected and utilized throughout the world as 'natural gas'. Methane, or 'marsh-gas' as it is sometimes called, is generated by the anaerobic (oxygen free) bacterial decomposition of organic matter and can be seen sometimes as bubbles rising to the surface of stagnant ponds. The methane often contains traces of phosphine PH_3 and diphosphine P_2H_4 which are also formed by similar bacterial action, the diphosphine igniting spontaneously as it mixes with oxygen in the air, also igniting the methane to cause small explosions or popping noises as gas bubbles to the surface of stagnant ponds, and also causing the flickering lights over marshy areas that are sometimes known as 'will o' the wisp'. This spontaneous ignition is probably the cause of the extensive slow fires that sometimes destroy large areas of peat bog and similar land when weather conditions have made it unusually dry.

Methane from these natural sources is not usually a danger in dwellings, probably because they are not constructed on soils with high methane emissions, but dangers can arise if structures act as collectors of methane from a wide area. The most tragic accident of this type involved an underground water pumping station in which methane was conducted along pipes and accumulated, causing an explosion and deaths and injuries amongst a party of visitors. Incidents involving dwellings are usually associated with construction on land fill and methane generation, sometimes on a massive scale, through bacterial decomposition of buried rubbish, this decomposition also causing serious subsidence damage. Although ignition of methane accumulations in fill is possible, and today fill at risk is usually carefully vented and monitored, the fill is open to the air and danger only arises through accumulations of explosive mixtures of methane and air in buildings where there are so many sources of accidental ignition. In 1986 a house at Loscoe in Derbyshire was completely destroyed in this way, despite the fact that it was not constructed immediately on top of the rubbish tip that was the source of the methane which caused the explosion, clearly demonstrating the danger of methane tracking through pipe trenches, and inadequately sealed entries into buildings.

The main precautions needed are a continuous vapour barrier to reduce the risk of methane diffusion into a building, coupled with adequate ventilation to disperse any methane accumulations that may arise despite the vapour barrier, precisely the same precautions that are described in section 18.8 to avoid radon dangers. Properly linked and continuous damp-proof membranes in solid floors and damp-proof courses in walls will provide efficient barriers to methane diffusion, provided that interruptions such as movement joints and service entries are properly sealed. With suspended floors it is theoretically possible to provide similar oversite protection, but discontinuities are likely to occur at walls and it is therefore best to ensure that the sub-floor space is well ventilated to disperse any leakage. For this reason it is probably best to avoid the use of suspended floors in high risk areas. In existing buildings a concrete slab will provide a good barrier to diffusion but cracks, movement joints, service entries and other openings to the soil must be carefully sealed. Accommodation ventilation may vary widely and in older buildings it may be as high as 4 ACH (air changes per hour), even with the doors and windows closed. In modern buildings without flues and with efficient door and window seals, rates as low as 0.1 ACH may occur, obviously too low to disperse any gas accumulation. In fact, bedroom ventilation rates lower than 0.5 ACH and living area rates lower than 1.0 ACH usually lead to 'stuffiness' and discomfort, and much higher ventilation rates are essential for gas cooking and flued combustion systems. It is only in extreme cases that ventilation rates may be too low for adequate methane gas control in conjunction with a vapour barrier; ventilation rates are significant only if a reliable vapour barrier cannot be provided, particularly in remedial works in existing buildings.

18.5 Legionnaires' disease

The need to eliminate pathogenic micro-organisms from water supplies has been generally recognised since 1854, when John Snow established that infectious disease can be transmitted through drinking water contaminated by sewerage, the death of Prince Albert in 1861 from typhoid fever prompting the rapid introduction of water purification schemes. Water is normally treated by filtration and chlorination, preventing any risk from drinking mains water, although water-borne infections still occur and are increasing due to changes in the use of water.

Although drinking water is normally supplied direct from the high-pressure main supply, hot water is often supplied through a low-pressure system involving a cistern or header tank, the cold supply to baths, basins and showers usually being taken from the same low-pressure system to ensure reliable temperature control with tap and shower mixers. This system does not involve any risk with reasonable use as chlorine in the water supply is sufficient to keep the header tank free from harmful organisms, but if the system is not used for a period, such as in a hotel which is closed during the slack season, organisms may develop in the header tanks, as well as sometimes in leaking taps and shower heads, which may be harmful, particularly through drinking the water during teeth cleaning or showering.

This problem attracted particular attention in July 1976 when delegates attending an American Legion convention in Philadelphia were affected by a mystery disease, subsequently traced to a bacterial infection of the water supply caused by an unknown bacterium which was later named *Legionella pneumophila*. Legionnaires' disease now accounts for about 2% of pneumonia in Britain, with about 200 cases diagnosed annually and a fatality rate of about 10%, men being three times more likely to develop the disease than women and those particularly at risk being between 40 and 60 years old. However, Legionnaires' disease is not commonly caused in Britain through infected low-pressure water supplies, but mainly through open cooling towers which are widely used in air conditioning plants for large buildings. In these cooling towers the hot water percolates over a fill and is cooled by a current of air induced by a fan. Droplets of water are often blown into the atmosphere in this way, causing infection over a wide area if the tower becomes infected. The risk is now very high because of the extensive use of these cooling towers; there are more than 200 installed in the Westminster area of London alone, and there have already been a series of major infections in London involving fatalities. Open cooling towers of this type are safe if the water is regularly treated to prevent infection, although the danger could be completely eliminated by the adoption of alternative cooling systems in which the water is not exposed to the air.

18.6 Allergies

Allergic reactions can be very complex and can be extremely dangerous if a person develops sensitisation to a stimulant, that is when initial exposure sensitises the individual so that subsequent exposure generates a massive reaction. Allergic reactions generally involve irritation of the skin, respiratory tract or eyes, usually through chemical stimulation, but sometimes as a reaction to particles of a particular size or shape.

Most people react to excessive exposure to dust, but persons are only considered to be allergic if they react unusually severely to particular stimulants. Carpenters and joiners are sometimes forced to give up work as they develop sensitisation to particular wood dusts, and failure to recognise sensitisation can result in very severe and even fatal respiratory reactions. Such problems are not usually encountered by occupants of buildings but they may be exposed to various other stimulants. For example, many people are very sensitive to the musty smell of a damp building, usually because they react to certain fungal spores. The Dry rot fungus *Serpula lacrymans* is particularly troublesome in this respect and is recognised as a major cause of asthma.

Reactions to volatile components in organic solvent wood preservatives are sometimes reported. Operatives applying remedial treatment by spray are obviously exposed extremely severely to these components and it is clear that they are not generally harmful from the excellent record of health in the industry. Mandatory controls have only been introduced comparatively recently, but all major manufacturers had accepted voluntary controls for many years earlier under the Pesticides Safety Precautions Scheme, manufacturing and labelling preservatives only in accordance with guidelines set by the Health and Safety Executive, and there is no evidence that wood preservatives are unacceptably toxic. Some complaints certainly have a psychosomatic origin, essentially fear generated simply by the odour of the preservative involved, but some complaints are justified. The usual cause is excessive application of preservative, perhaps deliberately in the belief that a more effective treatment is being achieved, although usually carelessness is the true explanation. A typical problem is preservative accumulated on oversites beneath floors through excessive treatment, releasing fumes over a very protracted period instead of the volatile components being lost rapidly as intended.

Extreme temperatures through spraying central heating pipes or through exceptional temperatures in roof spaces during strong sun can cause exceptional volatilisation of components, such as organotin and chlorophenol compounds, which are only slightly volatile in normal conditions, but most problems are certainly suffered by unusually sensitive individuals. Reactions to organotin compounds are generally associated with fair skinned individuals who also react strongly to any solvent, but some individuals seem to be sensitive to the musty odour produced by naphthenates and particularly the slightly sickly odour of the acypetacs compounds which have now largely replaced naphthenates, with other persons being particularly sensitive to contact insecticides such as Lindane. When problems occur through excessive application of preservative, the building can often only be normally occupied if materials treated with the preservative are removed; in the case of a floor treatment it is usually sufficient to replace the boarding alone in order to reduce odour to a tolerable level, even though the joists and oversite beneath may also have received treatment.

18.7 Arsenic dangers

In modern buildings arsenic occurs mainly as a component in water-borne copper-chromium-arsenic wood preservatives but fixation to the wood ensures that it does not present a health hazard in this form, except to a certain extent through dust caused during working treated wood. The most serious danger is not, in fact, direct poisoning by arsenic but accidental exposure to arsine gas formed from arsenic.

Arsine AsH_3 is also known as arsenic trihydride, arseniuretted hydrogen or hydrogen arsenide. It is a colourless, flammable and highly toxic gas with a mild garlic odour, although odourless at the threshold limit value of maximum safe concentration of 0.05 ppm in air. Acute poisoning causes headache, nausea, abdominal cramps and vomiting, progressing in more severe cases to massive destruction of red blood corpuscles causing haemoglobin in urine, jaundice and acute anaemia, followed by delirium, coma and death by kidney or liver failure. Repeated exposure to very low concentrations of arsine can cause anaemia, often in circumstances in which the true cause is not readily recognised.

Arsine is most frequently encountered today in industrial processes through the reaction of nascent hydrogen with arsenic, or water with arsenides of alkali metals, such as calcium arsenide, but in buildings arsine can be formed by the reaction of certain tolerant fungi, particularly *Scopulariopsis brevicaulis*, also known as *Penicillium brevicaule*, and *Paecilomyces* species, on arsenic compounds. This danger does not arise with properly designed copper-chromium-arsenic wood preservatives, such as those conforming with British Standard BS 4072 *Wood preservation by means of copper-chromium-arsenic compositions*, as the fungicidal properties of the copper prevents the establishment of these troublesome fungi. However, the main use for arsenic in buildings in the past was as Paris green and Scheele's green, arsenical green pigments which were particularly popular at one time in printing inks for wallpapers. If old wallpaper suffers condensation, rising or penetrating dampness, *Scopulariopsis brevicaulis* may develop, attracted by the protein content in the old horse-hoof size that was used, the fungus acquiring sufficient energy in this way to enable it to detoxify any copper in the pigment by conversion to copper oxalate and convert the arsenic content to arsine gas which will affect persons occupying the accommodation unless the ventilation is unusually generous.

The best known incident of arsine poisoning in recent years involved the United States Ambassadress in Rome who complained of disturbed nights and severe headaches. It was eventually realized that these symptoms might be the result of arsine poisoning caused by mould growth on green pigmented wallpaper. This incident drew attention to this danger and a health inspector visiting a restaurant in an historic Edinburgh building wondered whether arsine might be similarly generated by mould growth on old green pigmented wallpaper. A sample of the paper was analysed in the author's laboratory and found to contain copper arsenite; the mould was identified as *Scopulariopsis brevicaulis* which was apparently utilising food residues which had accumulated on the walls. In view of the danger of arsine generation it was recommended that the source of the dampness causing the mould growth should be identified and remedied, and the wallpaper should be stripped and replaced. This advice was not popular with the local historic buildings officer or the Edinburgh New Town Conservation Committee, who considered that the wallpaper was a valuable feature of the Georgian building which should be retained! The only realistic alternative in these circumstances was to emphasise the need to identify and remedy the source of dampness, and to use an effective and safe fungicide to control the mould. This is not, in fact, as simple as it might seem. The fungus concerned is a mould able to tolerate many fungicides, as is evident from its ability to break down Paris green containing both copper and arsenic, and many of the fungicides that are used in buildings tend to give poor control of moulds or are otherwise unsuitable because they are staining or volatile. The most suitable fungicides for controlling moulds in these circumstances are probably quaternary ammonium compounds, particularly the alkyl-benzyl-dimethyl ammonium compounds, such as benzalkonium chloride, which are simplest to use; they are supplied as 50% solutions which can be diluted to normal use concentrations of 1–3% with water.

Arsine can be generated by the same mould fungi in various other ways. For example, the plasticisers in PVC sheet and reinforced fabric are affected by bacterial and fungal infections, and resistant plasticisers or biocidal additives must be used in PVC which is intended for use in situations in which these infections occur. One of the most popular biocides for this purpose is 10,10'oxybisphenoxyarsine (OBPA) which gives good control of organisms such as *Streptomyces rubrireticuli* and *Aspergillus niger* which usually cause PVC contamination and plasticiser deterioration, but low retentions can be broken down by resistant fungi such as *Scopulariopsis brevicaulis*. OBPA is lost slowly by volatilisation and retentions gradually decline until there is a danger of arsine generation should conditions occur in which arsenic resistant mould fungi can develop. PVC sheet and reinforced fabric are used for marquee coverings, awnings, swimming pool linings and furniture upholstery coverings, but all these materials are encountered in conditions in which active fungal growth is unlikely and the danger of arsine generation is minimal. However, it has been recently discovered by the author that *Scopulariopsis brevicaulis*, which is commonly found in the home infecting damp protein materials such as wool carpets and leather, also infects PVC cot mattresses in the area affected by the warmth and perspiration of the baby, destroying the plasticiser and causing the PVC to become brittle. It was appreciated that there is a danger of arsine generation in these circumstances if OBPA is used in the material as a preservative. In fact, investigations showed that OBPA was not usually used in cot mattresses, but it was found surprisingly that the fungus could similarly convert antimony and phosphorus into stibine and phosphine gases which are also extremely toxic; antimony was found in almost all mattresses as antimony trioxide fire retardant additive, and phosphorus was found in some as phosphate plasticiser which is also used to improve fire retardancy. These observations prompted an investigation into the generation of these gases in this way as a possible cause of sudden infant death.

18.8 Radon dangers

The dangers to health arising from exposure to ionising radiation from radioactive materials have only been recognised relatively recently, particularly as a result of injuries caused by the use of early nuclear weapons and emphasised since then by incidents such as the Chernobyl accident. Such incidents involve radioactive material originating from mineral sources, and it is not therefore surprising that we are continuously exposed to radiation from natural minerals in addition to radiation from space and from man-made sources such as X-ray equipment. Natural mineral radiation affects us mainly in the form of radon gas diffusing from the soil beneath buildings and to a lesser extent from building materials.

Two types of radiation may be encountered. Gamma (γ) radiation is electro-magnetic energy, similar to X-rays and radiated in the same way as heat and light, the differences between these forms of energy being only the frequency or wavelength of the radiation. Emission of electro-magnetic radiation in this way results in loss of energy by the emitting substance which is generally characterised by cooling; most substances will only emit such radiation if energy is supplied from an external source, such as the emission of light when an electric current is passed through a lamp filament or the emission of X-rays when a 'target' is bombarded with electrons. Gamma radiation is sometimes emitted by radioactive substances when energy is supplied by 'decay' processes, in which a substance is converted from one nuclide to another by the second type of radiation involving the emission of particles from the nucleus. A nuclide is a particular isotope of an element; an element is characterised by its behaviour which depends upon its atomic electron configuration, whilst different isotopes of an element have the same electron configuration but different atomic mass. Thus in the uranium-238 decay scheme illustrated in part in Table 18.1, radiation of alpha (α) particles causes radium-226 to decay to radon-222 to polonium-218 to lead-214, and beta (β) particle radiation then causes lead-214 to decay to bismuth-214 to polonium-214; the element name in each case indicates the electron configuration and properties, whilst the number indicates the atomic weight and thus the particular isotope of that element which is involved. Chemical symbols are commonly used, so that radon-222 is often written as Rn-222 or ^{222}Rn.

The element radon, chemical symbol Rn and atomic number 86, is a gas in Group O of the periodic table, which also includes other noble or inert gases such as helium and neon. The element radon occurs as three isotopes, radon-219, -220 and -222, which are formed during the radioactive decay of actinium, thorium and

Table 18.1 Formation of radon-222 and its decay products

Radioactive decay of uranium-238 to lead-204 to show decay of radium-226 to radon-222, and formation of radon-222 decay products RaA, RaB, etc. (Prime 1987)

Nuclide	Half-life	Radiation
Uranium-238	4 500 000 000 years	alpha, gamma
leading to		
Radium-226	1622 years	alpha
Radon-222	3.8 days	alpha
Polonium-218 (RaA)	3.05 minutes	alpha
Lead-214 (RaB)	26.8 minutes	beta, gamma
Bismuth-214 (RaC)	19.7 minutes	beta
Polonium-214 (RaC′)	0.000 164 seconds	alpha
Lead-210	22 years	alpha
leading to		
Lead-204	infinite (stable)	–

uranium respectively, these isotopes being commonly known as actinon, thoron and radon. It is this latter radon (correctly radon-222) which is now recognised as the main source of radiation danger in buildings; thoron also contributes slightly but actinon is insignificant.

Radon-222 is a colourless, odourless and tasteless gas which decays through radiation to yield a series of decay products which are often commonly described as radon 'daughters' and designated RaA, RaB, etc., although they also each have nuclide or elemental isotopic nomenclatures, so that RaB is lead-214, as shown in Table 18.1. Although radon-222 is a gas, these decay products are all solids which become attached to particles and droplets in air as they are formed, settling in buildings and accumulating within the lungs. The radon-222 gas concentration in the air in a building can be minimised by ventilation but, whilst this ventilation reduces the concentration of the 'parent' from which these decay products or 'daughters' are derived, it does not remove any decay products that have already been deposited within the building or the body. Radon-222 is therefore important as the source of these decay products, but radon-222 is itself derived from radium-226 and ultimately from uranium-238 as shown in Table 18.1. Radon decay products therefore occur at highest concentrations in association with rocks containing unusually high concentrations of uranium, particularly in certain igneous rocks in, for example, south-west England and north-east Scotland. Mining represents the greatest hazard, but a significant risk can arise in buildings constructed on such rocks or, to a lesser extent, constructed from them.

The stability of each stage or nuclide in a radioactive decay sequence is indicated by its half-life, that is the time for half of the material to be lost by decay, as shown in Table 18.1. A relatively stable nuclide will have a long half-life, but an unstable nuclide will have a short half-life and will emit high intensity radiation because of the rapidity of the decay. It is apparent from the table that the radon decay products or 'daughters' have short half-lives, thus indicating higher intensity radiation than from the radon-222 gas 'parent', so that occurrence of this gas is not too serious, provided that its concentration can be minimised by ventilation in order to avoid as far as possible the formation of the solid and intensely radiating radon decay products which cannot themselves be removed by ventilation.

The significance of radiation in relation to health depends upon the intensity of the radiation and the period of exposure, producing 'acute' and 'chronic' symptoms. Radiation intensity is not significant for the low levels of radiation that arise in buildings from natural mineral sources, but radiation has an accumulative effect and it is the total dose, or the product of the intensity of the radiation and the period of

Table 18.2 Lifetime risks of premature death

Cause	Lifetime risk
All malignant neoplasms	0.2
Lung cancer from 100 mSv per year (2000 Bq/m^3)[a]	0.07 (0.2)
Lung cancer from 10 cigarettes per day	0.06
Lung cancer from 50 mSv per year (1000 Bq/m^3)[a]	0.04 (0.1)
Lung cancer from 25 mSv per year (500 Bq/m^3)[a]	0.02 (0.05)
Accidents in the home	0.009
Accidents on the road	0.008
Lung cancer from 5 mSv per year (100 Bq/m^3)[a]	0.004 (0.01)
Accidents to men at work	0.002
Cancer from nuclear waste[b]	0.000 001

Notes
[a] Exposure to radon-222 decay products only.
[b] Average public exposure 0.002 mSv per year.
A lifetime risk of 0.2 means that a person has a 1 in 5 chance of dying from this cause.
The original figures were based on International Commission on Radiological Protection guidelines that the risk of serious health effects such as fatal cancers and serious hereditary defects is 0.0165 (1 in 60) per Sv, suggesting a risk of fatal lung cancer of about 0.01 (1 in 100) from exposure to 10 mSv per year. This assessment of risk was revised in 1987–89 following research which indicated that the risks might be two or three times greater, as indicated by the figures in brackets, and this change prompted a reduction in action levels in the UK in early 1990.

exposure, that is important in relation to health. Radiation is usually measured in sieverts (Sv) per year. The National Radiological Protection Board (NRPB) originally recommended that the action level for remedial works to existing buildings should be 20 mSv per year and new buildings should be designed to limit exposure to below 5 mSv per year (NRPB 1987). These action levels were adopted in 1987 following recommendations by the International Commission on Radiological Protection, but the NRPB has since recommended that the action level at which remedial works are required in existing buildings should be reduced to 10 mSv per year and this recommendation was adopted by the UK Government in early 1990. These recommendations are based on assessments of the risk of fatal lung cancer due to this radiation in relation to other threats to health arising in dwellings. The lifetime risks of premature death through radiation and other causes are summarised in Table 18.2, but recent studies have shown that the risks are rather greater than originally assessed and this is the reason why action levels for existing buildings have been reduced (Wrixon 1985; Stather *et al.* 1988).

The lung cancer risk from exposure to radon-222 decay products in Table 18.2 must be related to total exposure to natural radiation. Table 18.3 shows that radon-222 decay products represent the largest single source of radiation, even at their average level of 0.7 mSv per year. In practical terms it is normal to measure the concentration of radon-222 that is present and to deduce the exposure to decay products that this represents, the NRPB action levels of 10 and 5 mSv per year in existing and new buildings being represented by radon-222 concentrations of about 200 and 100 Bq/m^3 (becquerels per cubic metre); these are the units and critical concentrations that must be used in considering radiation risks in buildings. National surveys, as well as more detailed surveys in high risk geological areas, are being undertaken by the NRPB, but even early results have demonstrated the wide range of radiation levels that may be encountered in dwellings through geological variations in different parts of the United Kingdom, as shown in Table 18.4 (O'Riordan *et al.* 1987; Wrixon *et al.* 1987). These surveys have also made it possible to estimate the number of dwellings in which radiation is likely to exceed the action level which may therefore require remedial works to reduce the health risks. It was estimated that about 30 000 buildings exceeded the original action level of 400 Bq/m^3, mainly

Table 18.3 Radiation from natural sources (NRPB data)

Source	Average annual dose (mSv)
Cosmic rays	0.30 (16%)
Terrestrial gamma rays	0.40 (21%)
Radon-222 (radon) decay products	0.70 (37%)
Radon-220 (thoron) decay products	0.10 (5%)
Other radiation	0.37 (20%)
Total average annual dose	1.87

Table 18.4 Surveys of dwellings for radon levels (NRPB data)

Region	Dwellings surveyed (Bq/m³)	Mean radon concentration (Bq/m³)	
		Living area	Bedrooms
National surveys			
UK	2000	28	20
London	123	13	11
Edinburgh	7	18	13
Cardiff	10	35	20
Belfast	3	12	9
SW England	50	80	63
E England	95	28	18
Aberdeen	9	18	10
Local surveys			
Cornwall	329	390	–
Devon	150	210	–
Swansea	7	14	–
Central uplands	150	180	–
Scotland	170	42	–

Notes
SW England includes Devon and Cornwall.
E England includes Norfolk and Suffolk.
Central uplands includes Derbyshire and Yorkshire.
Scotland includes Dumfries and Galloway, Grampian and Highland regions.

in Cornwall and Devon, but the reduction to $200 \, Bq/m^3$ in early 1990 increased the numbers of buildings at risk to about 90 000 and extended the need for monitoring to other areas.

The radon-222 decay products which represent the main radiation threat to health in dwellings are all formed from radon-222 gas. The main sources of radon-222 gas in buildings are the soil on which the building is constructed, the building materials, the water supply and the gas supply. Generally the soil in areas of uranium-rich minerals represents the main source. Building materials are much less significant, and the amounts of radon-222 likely to be introduced through water and gas supplies are less than the contribution from outside air, except when water supply is derived from uranium-rich sources or natural gas is being used close to its extraction point.

The main precaution that is necessary is to reduce the diffusion of radon-222 gas from the soil into the building accommodation. This requirement can be achieved most efficiently by ensuring that there is a continuous vapour barrier to isolate the accommodation from the soil, coupled with adequate ventilation of the accommodation. In practical terms normal damp-proof courses in walls and damp-proof membranes in solid floors will act as adequate vapour barriers, and ventilation must be provided for normal breathing, combustion and avoidance of condensation. However, damp-proof membranes must be continuous to provide reliable protection against radon-222 gas diffusion from the soil; movement joints and service entries must be sealed, and particular care taken to link membranes in solid floors with damp-proof courses in walls. With suspended floors it is theoretically possible to provide similar oversite protection, but discontinuities are likely to occur at walls and the sub-floor space should be well ventilated in order to disperse any leakage. For this reason it is probably best to avoid the use of suspended floors in high risk areas. In existing buildings a concrete slab will provide a good barrier to diffusion but cracks, movement joints, service entries and other openings to the soil must be carefully sealed to reduce radon-222 gas diffusion if tests have shown that radiation exceeds the action level of 10 mSv, equivalent to 200 Bq/m^3. Accommodation ventilation may vary widely and in older buildings it may be as high as 4 ACH (air changes per hour), even with the doors and windows closed. In modern buildings without flues and with efficient door and window seals rates as low as 0.1 ACH may occur. In fact, bedroom ventilation rates lower than 0.5 ACH and living area rates lower than 1.0 ACH usually lead to 'stuffiness' and discomfort, and much higher ventilation rates are essential for gas cooking and flued combustion systems. It is only in extreme cases that ventilation rates may be too low for adequate radon-222 gas control in conjunction with a vapour barrier; ventilation rates are significant only if a reliable vapour barrier cannot be provided, particularly in remedial works in existing buildings where electric extractors are sometimes necessary.

18.9 Asbestos dangers

Asbestos is the general name used for a group of amphibole or fibrous silicate minerals which includes crocidolite, amosite and chrysotile, as well as fibrous forms of actinolite, anthophyllite and tremolite. Asbestos has been extensively used as a fibrous reinforcement for cement in the manufacture of high density sheet materials, such as building boards and artificial 'slates', and with cement and other binders as low density fibrous insulation materials manufactured as sheets or formed by spray-gun application to pipes, tanks, boilers, undersides of roofs, structural steel work and wherever thermal insulation or fire protection is required. The asbestos used in the United Kingdom is mined mainly in Canada and South Africa, and in recent years it has been realised that it contains fibres which, because of their needle form, can cause irreversible lung damage and cancer or asbestosis.

Whilst asbestos fibres always represent a risk to health when present as dust in air, the most severe risks arise through exposure to fibres of crocidolite and amosite, commonly known as blue and brown asbestos respectively; in hazard to health terms, chrysotile or white asbestos is considered to represent a much lower risk and can be tolerated at $2\frac{1}{2}$ times the air concentrations of blue or brown asbestos.

The most serious risks are associated with working with asbestos products. Generally, a building contractor must take appropriate precautions when working with asbestos, and these precautions are most stringent when persons are exposed to blue or brown asbestos above critical action levels. A contractor has the option to prove by testing that persons will not be exposed to such levels of blue or brown asbestos, or alternatively adopt precautions as if these materials will be present. These requirements are described in detail in the Approved Code of Practice for *Work with asbestos installation, asbestos coating and asbestos insulation board*, revised by the Health and Safety Commission in March 1988, which a contractor must observe in order to comply with the requirements of the *Control of Asbestos at Work Regulations 1987*.

Occupants of a building containing undisturbed asbestos are not normally at risk, even if the asbestos is the most troublesome low density insulation form with blue or brown asbestos contents, as the critical fibres must be suspended as dust in the air in order to present a hazard to health. Asbestos fibres can usually only occur in accommodation air through asbestos work dust that has not been properly removed, or from air passing through or over insulation which is shedding fibres through breakdown of the binder. If it is suspected that asbestos fibres may be present in air, perhaps through breakdown of a sprayed asbestos roof insulation, samples of the insulation should be examined microscopically in order to check that it actually contains asbestos and to see whether blue or brown asbestos fibres are present. If these checks suggest that there may be a risk, air sampling should then be used to check whether the fibres are present in the air at levels that represent a significant health hazard. Whilst publicity in recent years has rightly emphasised the very serious health hazards that arise through exposure to asbestos, it has not been emphasised that it is dust in air that represents the main hazard and occupants in buildings are not normally at risk from undisturbed asbestos.

18.10 Sick building syndrome

Sick building syndrome is a condition in which the occupants of particular buildings suffer abnormal levels of sickness. Some of these symptoms are associated with 'wet' air conditioning systems and can be related to allergic reactions to spores or toxic reactions to bacteria, as previously described in section 18.5 of this chapter. Mites associated with carpets and soft furnishings may also cause allergic reactions. However, there are other symptoms that cannot be explained in this way. Headache and lethargy are sometimes reported in air conditioned buildings and can be related to very low carbon dioxide levels. Air conditioning equipment sometimes includes recirculation through scrubbers which are designed to reduce carbon dioxide levels, but any 'wet' recirculation system will function in this way. If carbon dioxide levels are too low, office workers and other sedentary persons will suffer from inadequate stimulus of respiration and low blood oxygen levels will then cause the observed symptoms.

In recent years a further series of symptoms have been recognised in sick building syndrome involving runny noses which become blocked, dry throats, thirst, tightness of the chest and difficulty in breathing, dry itchy eyes, perhaps with swelling and dry skin. These symptoms are actually caused by abnormally 'dry' air, that is air with a very low relative humidity. Modern air conditioning usually omits humidifiers to avoid 'wet' problems, particularly Legionaires' disease and some of the other microbiological problems previously described in this chapter in section 18.5. In cool winter weather the external air has a high relative humidity only due to the low temperature and it has a very low humidity or moisture content as explained in more detail in Chapter 4. When this air enters the building it is warmed, reducing the relative humidity and, if the humidity is already low through low exterior temperatures, the result is excessively low relative humidity in the building.

It can be seen from Figure 4.1 in Chapter 4 that, at a night temperature of 5 °C, air has a maximum moisture content of 5.4 g/kg (100% relative humidity), but at 20 °C in a building this represents a relative humidity of only about 37%, rather lower than the normal comfort level of 45–60%. However, many large buildings have mechanical ventilation which is excessive; a fresh air ventilation rate of 1 l/s per person is sufficient to ensure adequate oxygen supply and carbon dioxide clearance, but a minimum of 5 l/s is required by some codes with 8 l/s recommended, although

4 l/s is adequate with properly distributed ventilation to remove even body odours. In many modern buildings a fresh air ventilation rate of 12 l/s per person or more is used, a level appropriate to a conference or public room with heavy smoking, but many systems add the fresh air to recycled air, and total ventilation rates of 25 l/s are not unusual. These excessive ventilation rates result in 'wind chill' and temperatures must be increased in compensation to maintain comfort, typically to 23–25 °C. The effect of increasing the temperature from 20 °C to 25 °C in the previous example is to reduce the relative humidity in cold winter weather to about 27%, far below comfort level and the cause of the 'dry' sick building symptoms which are now such a problem in some buildings.

These problems are easily remedied by using only fresh air for ventilation, so that the ventilation rate can be substantially reduced, perhaps to levels of only about 6 l/s per person in normal office accommodation, a change that will avoid the wind chill and permit air temperatures to be reduced to about 20 °C, increasing relative humidity and avoiding 'dry' sick building symptoms except in very frosty weather. These changes also reduce energy consumption for heating and air circulation.

Plate 2. Stone erosion at St Paul's Cathedral, London. The Portland limestone was protected within the arch, but horizontal surfaces were slowly eroded where they were exposed to rainfall containing dissolved pollutant gases.

Plate 1. Frost damage affecting stone masonry at Lincoln Cathedral. The stone was normally durable but suffered frost damage where it was affected by rising dampness and therefore had a high moisture content in freezing conditions.

Plate 3. Stone erosion at St Andrew's Church, Glasgow. The carbonaceous sandstone columns absorbed acid rainwater which dissolved the carbonate binder and deposited calcium sulphate within the surface, stresses in the outer surface than causing it to separate from the weakened stone beneath.

Plate 4. Portland limestone erosion in Blackpool. Erosion of isolated cladding panels was due to pore size variation; the durable panels were predominantly macroporous but the eroded panels were microporous.

Plate 5. Stone incompatibility at Lincoln Cathedral.
The stone repair was very durable but also very porous, allowing rainwater to accumulate on the original stone beneath and cause frost spalling during freezing conditions. Similar fracture and erosion damage can be caused by absorption of limestone washings by a sandstone, or even magnesian limestone washings by a calcareous limestone.

Plate 6. Stone cleaning with biocides. Most of the dirt on stone results from biological activity. The clean headstone in this cemetery maintained by the Commonwealth War Graves Commission had been sprayed with a biocide which halted growth and resulted in weathering causing progressive natural cleaning. The Commission now uses biocides to maintain all headstones in clean condition.

Plate 7. Rainwater staining on concrete. The brickwork above the concrete absorbed rainwater but the glass shed it onto the concrete, causing wetter areas with heavier biological growth; unsightly contrasts can be reduced by applying biocide or water repellent treatments to the concrete.

Plate 8. Stone damage caused by crustose lichen growth. The lichen dissolves calcium carbonate and deposits it as calcium oxalate in the surface of the stone, the toughened surface then separating from the weakened stone beneath.

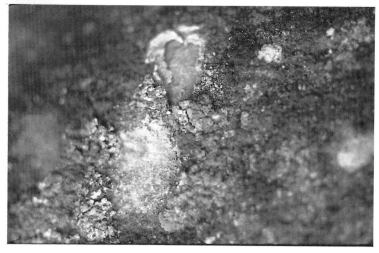

Plate 9. Metal corrosion damage to stone and concrete. Corrosion of ferrous dowels and cramps in masonry and steel reinforcement in concrete causes expansion and spalling damage.

Plate 10. Death Watch beetle damage. Death Watch beetle normally attacks only damp and slightly decayed hardwoods and it is therefore particularly associated with old buildings; a variety has developed in the Channel Islands which also attacks damp softwoods in modern buildings.

Plate 11. Powder Post beetle damage to a plywood floor. Powder Post *Lyctus* beetle attacks only fresh sapwood of certain hardwoods but it can cause complete destruction.

Plate 12. Bark borer damage on a softwood rafter. The Bark borer *Ernobius mollis* causes damage which is often confused with Common Furniture beetle, but Bark borer is confined to the bark and the wood just beneath it and remains active for only about two years after a tree is felled so that treatment is not required.

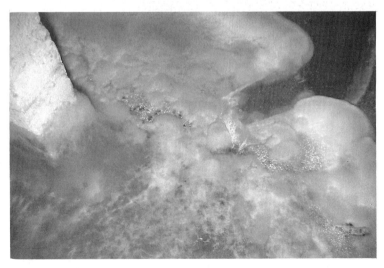

Plate 13. The Dry rot fungus. The Latin name *Serpula lacrymans* or weeping fungus refers to the 'tears' of water that develop on active growth, maintaining the humidity of the atmosphere.

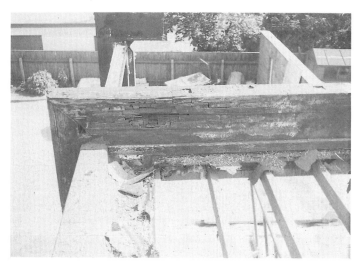

Plate 14. Fungal decay in modern buildings. This external structural frame first suffered Wet rot damage and it was covered with felt to prevent rain penetration, but Dry rot, stimulated by acidity from the original Wet rot attack, developed as the moisture content slowly decreased.

Plate 15. Fungal decay in modern buildings. Decay caused by interstitial condensation in the sheathing plywood on a timber frame house; condensation accumulated because the inner vapour barrier was damaged and the cavity between the sheathing plywood and the external brickwork cladding was unventilated. Decay due to condensation can also develop in floors laid on battens on concrete slabs if the floor space is not ventilated to the exterior; condensation is encouraged if the space is ventilated to the interior.

Plate 16. Fungal decay in joinery (millwork). Water absorption through end-grain surfaces can cause decay in untreated wood joinery, decay in susceptible wood often spreading to resistant wood.

Plate 17. Fungal decay in flat roofs. Interstitial condensation in roof decking often causes fungal decay damage; condensation can be avoided by ventilating the space between the decking and the ceiling insulation in order to disperse moisture diffusing from the accommodation.

Plate 19. Concrete texture. The colour variation in these concrete tiles was caused by poor control of water content during manufacture, the lighter colour being due to excessive water and cement laitance on the surface.

Plate 18. Heat loss through a flat roof. On this felted roof photographed after rainfall the light lines have been dried by heat loss through the metal framework supporting the cement-bonded wood wool insulating slabs.

Plate 20. Frost spalling in clay roof tiles. Frost spalling is caused by prolonged retention of water by microporous tiles which are more likely to be saturated in freezing conditions. Microporosity is often associated with light coloured underfired tiles; similar failures occur in clay bricks.

Plate 21. Water penetration in low pitch roofs. Low pitch roofs have poor resistance to rain penetration in windy conditions, even if closely bedded coverings are used such as the western red cedar shingles in this photograph; the sarking felt beneath the covering often gives temporary protection and problems may not develop for several years, usually at the eaves where the sarking felt deteriorates most rapidly.

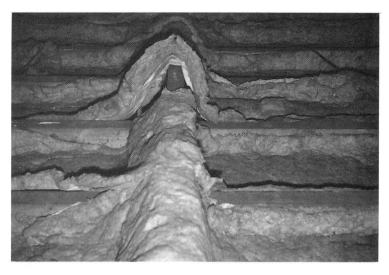

Plate 22. Remedial wood preservation problems. Ceiling insulation must be removed to allow treatment of ceiling joists; if organic solvent preservatives are applied to glass or mineral fibre insulation there is a danger of fire through auto-ignition.

Plate 23. Remedial wood preservation problems. If some organic solvent formulations are applied too generously, solvent vapours are absorbed by PVC cables, causing bleeding of the plasticizer which can stain ceilings and walls but also cause arcing and short circuits.

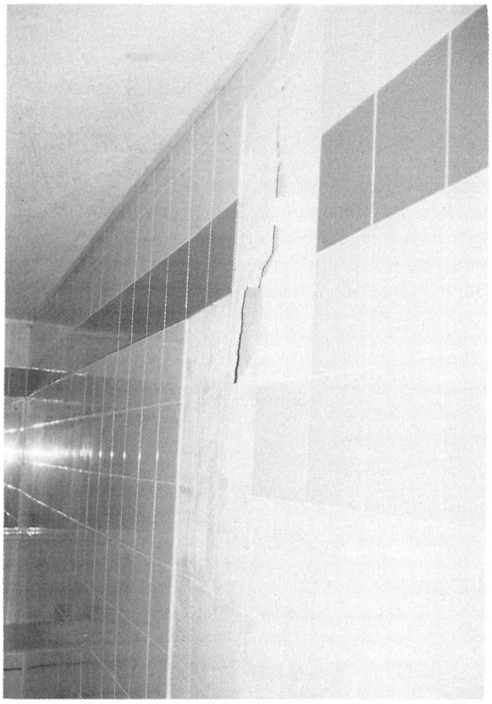

Plate 24. Damaged wall tiling. Wall tiles can be dislodged by differential movement, usually through shrinkage of cast concrete or concrete block backgrounds; normal good practice requires movement joints through the tiling to relieve these stresses.

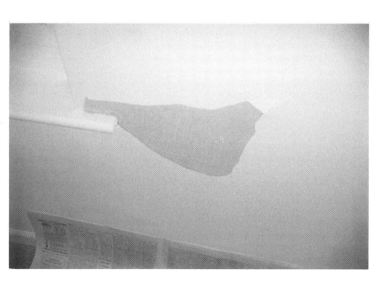

Plate 25. Plaster detachment. Gypsum plaster separation from a cement/sand undercoat in dry conditions is caused by a failure to wet the surface thoroughly before plastering.

Plate 26. Penetration of wood preservative.
Penetration depends on the properties of the
preservative and the wood, as well as the moisture
content of the wood; in sample 1 the sapwood is
completely penetrated, but in sample 11 sapstain
deterioration has increased porosity and allowed some
penetration into the heartwood, whilst only limited
penetration has been achieved in sample 2 because it
had a high moisture content and no space for the
preservative.

Plate 27. Corrosion of steel railings. Corrosion developed on these
steel railings shortly after erection and decoration; a zinc-rich primer
had been used to inhibit corrosion but it is only effective if applied to
clean steel, and in this case it was applied over mill scale, old red oxide
primer and rust.

Plate 28. Stone cladding fractures.
Fracturing of this granite cladding panel was
caused by failure to continue the horizontal
movement joint around the adjoining transom
feature; the transom granite cladding also
fractured.

Plate 29. Stone cladding fractures. Fracturing of
this granite cladding panel was due to expansion of
positioning dabs; gypsum plaster was used to thicken
the cement mix, water vapour absorption
subsequently causing sulphate attack of the cement,
expansion of the dabs and fractures of the cladding.

Plate 30. Iron stain on granite cladding. These patches of brown stain on light grey granite were caused by 'rust' formation following decomposition of iron pyrites inclusions during flaming to give a rough finish.

Plate 31. Fractures in concrete cladding. Fractures in concrete cladding are usually caused by reinforcement corrosion through inadequate cover, but in this case the panel was damaged during erection, partly because it was being handled too soon after casting and was rather weak.

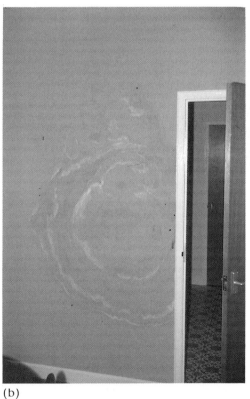

(a)   (b)

Plate 32. Condensation in a boiler flue. (a) Condensation, due to inadequate air flow in the flue, dripping from the flue pipe above the boiler; (b) Condensation and flue acids causing damage to decorations in the room above the boiler.

Plate 33. Water penetration in a new cast stone building. Cast stone is usually manufactured using a semi-dry mix to give an interesting surface texture but it is difficult to compact and is often very porous; water penetration into porous cast stone in this case caused flooding of the trays over windows and lower roof extensions with severe dampness internally.

Plate 34. Deterioration of GRP roof panels. Glass-reinforced polyester roof panels usually deteriorate steadily, exposing the glass reinforcement, but fire retardent panels deteriorate more rapidly with loss of the more soluble phosphate fire retardent which often encourages lichen growth if washing contaminates adjacent cement roofing panels.

Plate 35. Lime deposits on masonry. These heavy lime deposits were caused by water saturating a semi-dry bed under paving on the flat roof above, the lime solution then emerging from the masonry through failure to provide an edge upturn to the roof damp-proof membrane beneath the paving.

Plate 36. Sulphate efflorescence on bricks. Bricks containing excessive soluble sulphate often suffer from unsightly white efflorescence or crystallization, but the sulphate may react with the cement in the mortar joints to cause expansion and perhaps severe structural damage; sulphate resisting cement should always be used with bricks containing excessive sulphate.

Plate 37. Render fractures. The fractures in this render are caused by sulphate attack expansion of the cement in the mortar of the supporting brickwork; the bricks were known to contain excessive sulphate and sulphate resisting cement was specified but not used.

Plate 38. Paint deterioration in a marine environment. The electrophoretic acrylic coating on these new aluminium window frames was damaged by alkali attack; sea spray reacted with the lime in the adjacent new rendering to form sodium hydroxide.

Plate 39. Failure of epoxy resin flooring. This industrial floor was laid as a resin and aggregate base coat or screed followed by a resin finish, but the basecoat resin did not cure because it was not properly mixed or the temperature was too low; inadequate cure was indicated by a strong phenolic odour.

Plate 40. Blister formation in a swimming pool floor. The pool floor comprised a reinforced concrete slab and cement/sand screed, topped with an epoxy resin screed and finish, but the finish was not completely impermeable and the screed resin did not cure, producing concentrations of phenolic compounds which caused water flow by osmosis through the semi-permeable finish and development of the blisters.

Plate 41. Failure of polymer-modified cement flooring. This polymer-modified cement and aggregate floor topping was laid in a bottling hall without movement joints, stresses causing fracturing and separation from the concrete base, but the floor topping also 'domed' beneath the syrup machines through expansion caused by sugar reaction with cement; cement-based flooring must never be used if sugar contamination is possible.

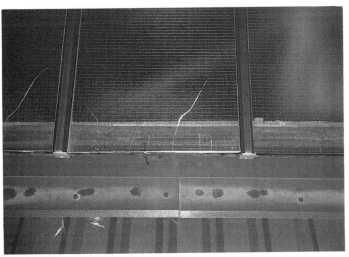

Plate 42. Fractures in patent double glazing. These industrial patent glazed roof lights were fitted with sealed double glazing units with inner wired glass and outer plain glass, the units projecting well beyond the seal so that, when the inner wired glass was expanded by the accommodation heat, the projecting cool glass was stressed, causing the observed fractures.

Plate 43. Condensation in sealed doubleglazing. Condensation within these double glazing units was due to defective seals; stepped units were involved and the defective seals allowed warm humid air from the accommodation to enter the units, causing condensation on the inner surface of the outer glass.

Plate 44. Sulphate attack in clay brickwork. This new brickwork to an octagonal building was exposed to heavy rainfall which caused sulphate in the bricks to migrate and react with cement in the mortar to cause expansion, increasing the height of the brickwork but also causing the observed distortion.

Plate 45. Limeblowing in clay bricks. Shells embedded in the brick clay were reduced to quicklime during firing, but slow absorption of moisture from the air caused expansion and rupture of the brick surface.

Plate 46. Differential movement between brickwork and wood. In dry weather shrinkage of the wood frame formed a gap which was then pointed with mortar, but subsequent expansion of the wood caused spalling of the adjacent brickwork.

Plate 47. Render failure due to water penetration. The damp-proof course beneath the concrete wall copings was trimmed against the edge of the brickwork before rendering so that it did not protect the top edge of the render; water penetrating through the coping joints accumulated behind the render which became detached during frost.

References and further reading

References

Boyer, J.T. (1988) *Guide to Domestic Building Surveys of Residential Properties*, Architectural Press, London.

Cutler, D.F. and Richardson, I.B.K. (1981) *Tree Roots and Buildings*, Construction Press/Longman, London.

Driscoll, R.M.C. (1983) 'The influence of vegetation on the swelling and shrinking of clay soils in Britain', *Geotechnique*, **33**(2) 93–105.

National Radiological Protection Board (1987) *Exposure to Radon Daughters in Dwellings*, National Radiological Protection Board Publication ASP10.

O'Riordan, M.C., James, A.C., Green, B.M.R. and Wrixon, A.D. (1987) *Exposure to Radon Daughters in Dwellings*, National Radiological Protection Board Publication GS6.

Prime, D. (1987) Exposure to radon decay products in dwellings, *J. Roy. Soc. of Health*, **6**.

Richardson, B.A. (1980) *Remedial Treatment of Buildings*.

Richardson, B.A. (1978) *Wood Preservation*.

RICS (1985) *Structural surveys of residential property* (rev. edn), RICS guidance note.

Stather, J.W., Muirhead, C.R., Edwards, A.A. *et al*. (1988) *Health effects models developed from the 1988 UNSCEAR report*, NRPB publication R226, HMSO, London.

Wrixon, A. (1985) 'Radon – natural health threat', *Surveyor*.

Wrixon, A.D., Green, B.M.R., Lomas, P.R., Miles, J.C.H., Cliff, K.D., Francis, E.A., Driscoll, C.M.H., James, A.C. and O'Riordan, M.C. (1987) *National Radiation Exposure in UK Dwellings*, National Radiological Protection Board Publication R190.

Further reading

Building Research Digests are the best source of further information and guidance, including references to current literature. Digests are published monthly by the Building Research Establishment, Building Research Station, Garston, Watford WD2 7JR.

British Standard specifications and codes of practice are also very helpful, as well as similar standards published in other countries.

The National Radiological Protection Board (Chilton, Didcot, Oxon OX11 0RQ) publication *Documents of the NRPB* summarises current formal advice of the Board and practical guidance; Vol. 1 No. 1 was issued in January 1990.

Index